WHY GREENLAND IS AN ISLAND, AUSTRALIA IS NOT— AND JAPAN IS UP FOR GRABS

WHY GREENLAND IS AN ISLAND, AUSTRALIA IS NOT— AND JAPAN IS UP FOR GRABS

A Simple Primer for Becoming a Geographical Know-It-All

Joyce Davis

QUILL

WILLIAM MORROW

NEW YORK

Library of Congress Cataloging-in-Publication Data

Davis, Joyce, 1950–
    Why Greenland is an island, Australia is not—and Japan is up for grabs : a simple primer for becoming a geographical know-it-all / Joyce Davis.
        p.   cm.
    ISBN 0-688-10176-3
    1. Geography.   I.   Title.
G116.D38   1993
910—dc20   93-20248
                                                           CIP

Printed in the United States of America

First Quill Edition

1 2 3 4 5 6 7 8 9 10

BOOK DESIGN BY MMDESIGN, 2000, INC./MICHAEL MENDELSOHN

To the geography students of
Grace Church School in Manhattan—
who have been a special inspiration to me
in my own study of the subject

ACKNOWLEDGMENT

I'm grateful to my good friend
William Proctor for his help in organizing
and shaping this manuscript.

LIST OF MAPS AND DIAGRAMS

CONTENTS

The World

Greenland

Germa

Canada

United
Kingdom

Ireland

France

Portugal

Spain

United States

A

Mexico

Central America

Venezuela

Colombia

Niger

Ecuador

Brazil

Peru

Bolivia

Paraguay

Uruguay

Chile

Argentina

Robinson Projection

andinavia

Poland

Russia

Czechoslovakia

Yugoslavia

Italy

reece

Japan

ya

China

India

Iraq

Sudan

Ethiopia

Zaire

Kenya

Tanzania

Indonesia

Madagascar

Australia

South
Africa

New
Zealand

Antarctica

CHAPTER ONE

WHY GREENLAND IS AN ISLAND, AUSTRALIA IS NOT—AND JAPAN IS UP FOR GRABS

We have all puzzled over geography at one time or another, and some of us may even have uttered confused or ill-considered comments like these:

"Sure I know a lot about Iraq as a result of the Persian Gulf War—but what are those countries on its northern border? And I'm not sure, but wasn't Persia the ancient empire that was located in present-day Iraq?"

"Now, Turkmenistan—is that a province of Turkey? Is it the same as Turkestan? And exactly where *is* Turkmenistan, by the way?"

"I fly through Chicago frequently—the second-largest city in the USA and the capital of Illinois, right?"

"What's the deal with Yugoslavia? As I recall, there are two main republics in the country, Croatia and Bosnia, and they declared their independence and then got into a fight with Serbs."

"I can name all five continents. But I never could understand why Greenland is considered an island rather than a continent—it's so much bigger than Australia!"

Of course, *none* of the above observations is completely correct. Here's a sampling of the errors:

- The nations on the northern borders of Iraq are Syria, Turkey, and parts of Iran. The ancient empires that ruled on the current site of Iraq included the *Assyrians* and *Babylonians,* but not the Persians (they were located where Iran is today).

- Turkmenistan, situated just northeast of Iran on the Caspian Sea, was one of the republics that comprised the Union of Soviet Socialist Republics. Turkmenistan declared its independence from the USSR on October 27, 1991.

  Furthermore, Turkmenistan is *not* the same as Turkestan (sometimes spelled Turkistan), which is a broad central Asian area that was divided earlier in this century between China, Afghanistan, and the old USSR. Turkmenistan is a part of the former region of Turkestan, but the two are not synonymous. (To confuse matters still further, Turkestan is also a city in the old USSR republic of Kazakhstan, which declared its independence on December 16, 1991.)

- Chicago is the *third*-largest city in the United States after New York and Los Angeles, and *Springfield* is the capital of Illinois.

- *Four* Yugoslavian republics declared their independence from Yugoslavia between 1991 and 1992: (1) Croatia; (2) Slovenia; (3) Macedonia; and (4) Bosnia-Herzegovina. The Yugoslavian republic of Serbia, main home of the Serbs, has opposed this drive for independence.

- There are *seven* continents, and Greenland is actually considerably *smaller* than Australia. It just looks larger because of the distortions that occur when the contours of the globe are reproduced on a flat map.

As I've been on junkets to many parts of the globe, I've been amazed at most people's limited knowledge of geography, even among cosmopolitan travelers. This should come as no surprise to me, given the recent surveys and studies that show how inadequate the understanding of geography is among adults, especially those in the United States.

In a Gallup poll conducted for the National Geographic Society, nearly eleven thousand people in nine developed countries were given a simple test asking that they identify sixteen regions in the world. There were no trick questions; they just had to match numbered areas on a world map with the names of major countries or bodies of water.

Among all adults, the United States ended up in a lackluster sixth place ahead of the United Kingdom, Italy, and Mexico. But in the 18–24 age group, the US came in *last*, with an average of only 6.9 right answers out of 16. An incredible 14 percent of Americans couldn't even pick out the United States on the map! The *best* national scores in the 18–24-year-old group (registered by those in Sweden and West Germany) weren't that great either, with an average of fewer than twelve right answers out of sixteen.

Why is our knowledge of geography so abysmal?

In the first place, the subject isn't emphasized in our schools. Seeing this problem, I took steps a number of years ago to establish a comprehensive geography curriculum at Grace Church School, a nonsectarian private school in Manhattan. I also organized an annual geography bee, in conjunction with a program sponsored by the National Geographic Society. The interest in geography—and the expertise of the average student—soared after these simple steps.

In the second place, after leaving school, people fail to realize how essential a basic grasp of geography can be in furthering their careers. This is especially true for those who have to foster overseas contacts or interact with foreign clients or colleagues.

As a native of the United Kingdom, I know how small the world can be when you begin to travel in Europe, Asia, the Middle East, or other parts of the globe. Although many Americans may never leave their native country, business people and professionals from Europe, the Middle East, and the Far East are more accustomed to moving back and forth between different countries and cultures. They expect those from other lands to know something about their homeland and customs—and a lack of such knowledge may be interpreted as either poor education or a disdain for them as people.

WHAT EXACTLY IS GEOGRAPHY, ANYWAY?

To ensure that we begin this discussion on the same footing, I want to see that we have a common understanding of exactly what geography is all about. Geography is the study of the earth's surface, and all the movements of land, sea, and air that affect the earth. But it's also the study of our relationships with other people and places.

To show you what I mean, take a close look at yourself first thing in the morning. You may be awakened by an alarm clock made in Japan, and your first meal may consist of coffee from Colombia, bacon from Canada, and orange juice from Florida. And don't forget the bread and cereal from the Great Plains. You may then jump into a German car, while wearing an Italian knit sweater and English slacks. If it's hot, you might wear instead a Madras shirt from India and an Irish linen jacket.

Have you ever stopped to think: Where are all these places that are enhancing my life? What are these countries like? Who lives there—and how do they make and export the items that I'm using? If you're interested in asking questions like these, you have the makings of a *real* geographer.

The first geographers were the Greeks more than two thousand years ago. In fact, geography is a Greek word that means "writing about the earth." It emphasizes *where* things are, as well as the human and environmental characteristics of the world. In these modern times, one of the most important aspects of geography is the impact that human activities have on the environment. If we pollute the planet Earth with toxic wastes, deplete its natural resources, overpopulate it, or in other ways abuse it, there may be consequences that can start a disastrous, far-reaching chain reaction.

The more we learn about geography—where we live, where and how others live, what our similarities and differences are, and how we can alter the future of the earth—the greater our capacity will be to influence our future for the better.

HOW TO BECOME A GEOGRAPHICAL VIRTUOSO—IN JUST A FEW MINUTES A DAY

This book is designed to *highlight* the most important geographical issues you're likely to encounter in everyday life. The emphasis is on ease and speed of assimilating this information. We'll occasionally take a pressing topical issue—such as the 1991 Persian Gulf

War, the Yugoslavian conflict, and starvation in Somalia—and show how an understanding of geography can strengthen your grasp of the situation.

After reading through the following pages just once, you may even find yourself becoming something of a geographical virtuoso—at least in comparison with 99 percent of the other people you encounter at work or in your social activities.

To give you an idea about how easy it is to expand your knowledge of geography, consider the following simple comparison of continents and islands—with Greenland, Australia, and Japan as case studies.

WHAT ABOUT GREENLAND, AUSTRALIA, AND JAPAN?

A continent would be defined by most geographers as one of the seven large, continuous segments of Earth into which the land surface is divided. A continent may be surrounded entirely by water or may be linked by land to one of the other continents.

In contrast, an island is an area of land that is completely surrounded by water and is *smaller* than a continent. How much smaller? By fiat of the geographical powers-that-be, Australia is as small as you can get and still be an island continent.

Yet I've heard so many of my students, as well as plenty of adults, protest: "But Australia looks smaller than Greenland on my map!"

It's quite true that on many flat maps, Greenland may look *much* larger than Australia. In fact, however, the area of Australia is nearly three million square miles, while the area of Greenland is only eight hundred forty thousand square miles. On the standard Mercator flat-map projection (used since the sixteenth century by navigators), Greenland also seems to be nearly twice the size of South America—even though it's actually only a little bigger than Mexico.

Why the disparity between the maps and reality? When the land on a globe is projected onto a flat map, there is always a major distortion of areas at latitudes near the poles when compared with land masses closer to the equator. As a result, Greenland, which is near the North Pole, appears on a flat map to be considerably larger than it really is.

When you apply these continent-island distinctions to an area like Japan, it may seem clear that Japan is an island—or does it? Actually, Japan is not an island but an *archipelago,* or a group or cluster of islands. The main islands that make up Japan are Honshu (the largest island that contains Tokyo), Hokkaido, Kyushu, and Shikoku.

But there's another problem with Japan—a threatening geographical fact that causes me to say this archipelago nation is "up for grabs." You see, Japan sits on the Pacific "Ring of Fire," the horseshoe-shaped volcanic zone that many experts say runs from the Aleutian Islands, to southeast Asia, to the southern tip of South America. Others trace the limits of the Ring according to the earthquake-prone areas that are bounded on the west by Japan and on the east by California, with its San Andreas Fault. Over the millennia, the "crustal instability" of this region has caused parts of Japan to sink into the ocean or otherwise change drastically. With the presence of about forty active volcanoes and a history of devastating earthquakes, the nation's future geographical status is uncertain indeed.

As you can see from this brief geographical excursion, a simple comparison of two or three parts of the world can reveal a gold mine of information about the past, present, and possible future of Earth. Certainly, anyone who does business in the far-flung lands of the Pacific Ocean should be familiar with these basic facts about Japan and Australia. An Australian or Japanese colleague might not expect you to know about the different Japanese islands, the relative size of Australia, or the Ring of Fire—but you can bet that he or she will be pleasantly surprised and impressed if you do.

The remaining chapters have been set up to provide you with practical situations or comparisons, such as the foregoing Greenland-Australia-Japan illustration. The objective is to provide tools that you can *use* in your daily life and work—as you'll see in this brief guide to using the book.

HOW TO USE THIS BOOK

There's enough information in the first two chapters to give you the approach you need to discuss and deal intelligently with current events containing an important geographical component.

As you move step-by-step through this book, you'll gradually build an ever-widening base of important information about the earth, the nations, and the cities, including:

a formula to gain a grasp of the geographical implications of current events in a matter of seconds

memory devices to remember different natural and man-made features of the earth

the twenty most important world cities—and techniques for remembering an even larger number of urban centers and capitals in the United States

the characteristics of straits, isthmuses, peninsulas—and various other features of our lands and seas

how to read different types of maps

basic facts about global divisions, such as the hemispheres and time zones

the "language of location," or the special geographical terminology that every well-educated person should know

At the end of the book, I'll include some quizzes so that you can test your growing knowledge of geography and also review what you've learned.

Now, let's take a closer look at some present-day lands that are in considerable political or economic turmoil—and see how geography can enhance your understanding of their difficulties.

CHAPTER TWO

TERRITORIES IN TURMOIL

An understanding of geography is an absolutely integral part of being able to read and discuss current events intelligently.

To illustrate how geography plays such a role, I'm going to identify several parts of the world that are in turmoil as I write—and that I expect will continue to be in some degree of uncertainty. But the key thing you should focus on in the following sections is *not* the substantive political, economic, or historical events. After all, times and circumstances do change. Rather, I want you to approach these situations as *models* for your own reading. The emphasis will be on how to use the geography, or the physical location and characteristics, as a tool to grasp what's going on in particular countries and regions.

I've chosen two models:

the vast region covered by the old Union of Soviet Socialist Republics (USSR)

the Middle East, with a particular emphasis on Iraq

But before we begin to examine these specific geographic situations, let's first formulate a simple "geographical strategy," which should enable you to analyze the locus and physical features of any event you encounter.

YOUR PERSONAL GEOGRAPHICAL STRATEGY—FOR MORE INTELLIGENT READING AND DISCUSSION

Suppose you're reading an article or passage in a book; you're watching a television program; or you're expecting to participate in a discussion that has some geographical dimension. If you first take the following steps, you'll greatly enhance your understanding and enjoyment:

Step #1: Immediately identify the geographical issue.

It's usually impossible to understand a complex war or political situation without first knowing something about the geography of the region. You have to develop a grasp of the boundaries, and the physical and historical characteristics of the nations, republics, or regions involved. For that matter, if you're scheduled to discuss a marketing proposal for certain cities that are served by your corporation, your first step should be to identify those particular areas on a map. That way, you'll have a framework for doing further research.

Step #2: Study closely all maps reproduced in or accompanying the reading material you're using.

Often, a difficult news story can become quite clear after you spend a few seconds looking at an accompanying map. Or if you're traveling to an unfamiliar region or country, spending some time with maps of the area is absolutely necessary to give you a fast "feel" for the new location. Without this sense of geography, you'll even lack the ability to move about with some degree of ease.

You may even be given a map related to some issue concerned with your business or with a personal interest. In these cases, a geographical analysis can be invaluable in helping you develop a "big picture," or broad perspective of the issue.

Step #3: Compare the maps you're given with more detailed maps in an atlas.

Most of the time, you won't be given a detailed map when you're reading a news account or some other written material. To acquire a more precise view of the area, pull out an atlas and check some of the towns and physical features that may not have been included in the original depiction. As part of this exercise, you may want to look at a topographical map showing the peculiar terrain, including mountains, valleys, and specific elevations.

Step #4: Compare the maps you're given with large-area *maps showing how the area in question fits into a broader geographical context.*

The map in your reading material or on the TV screen will probably provide only a limited view of how the region fits into surrounding areas. So again, pull out an atlas and note how the particular area fits into surrounding nations and regions.

Step #5: Combine what you now know about the geography with the other facts involved in the situation.

If you're reading about some political or military problem, use your imagination to place the actors in that drama where they belong on the map. As much as possible, tie in the other facts to specific geographical sites.

Step #6: Close your eyes and try to picture the scene you've been studying.

This is a way of testing yourself to see how much you remember and understand about the geography and other features of the area you're reading about or studying. If you can bring to mind all the main physical features—including what you may have learned about population, economic features, and political events—you can be relatively sure that you've mastered the main outlines of the geography of your subject. And you can also be sure that your understanding of the topic is light years ahead of those who fail to take geography into consideration.

Now, with this simple strategy in mind, let's move on and try applying it to our first specific situation: the breakup of the old USSR.

WHAT EXACTLY HAPPENED TO THE USSR?

It's impossible to understand what happened to the Union of Soviet Socialist Republics in 1991–1992 without focusing on geography. Below, I've included a brief scenario of what happened, with the primary emphasis on the *physical* fragmentation of the Soviet empire—a fragmentation that promises to continue exerting a major influence on world affairs well into the future.

As you read through this sequence of events, apply the six-step strategy you learned in the previous section. Use an atlas, as well as the sectional maps provided in the text. Above all, to develop a personal, internal understanding of what happened, picture in your mind's eye how this great Soviet empire fell apart, piece by piece, over the course of just a few months.

This scenario, presented in chronological order, should provide you with a framework to move through Steps #1 and #2 in the geographical strategy. In other words, you'll have the tools to identify the important sites and regions on a map. As you proceed, use the maps at your disposal to make your identifications of the italicized nations and regions.

Event One: *Lithuania* makes a declaration of independence from the USSR on March 11, 1990. But the USSR establishes an economic blockade and fourteen are killed when Soviet paratroopers move into the country. The declaration of independence is suspended.

Lithuania is one of the *Baltic* states just to the east of Russia on the Baltic Sea. (The others are *Latvia* and *Estonia.* The Baltics, by the way, shouldn't be confused with the *Balkans,* or the nations situated on the Balkan Peninsula in southeast Europe, between the Adriatic and Ionian seas on the west and the Aegean and Black seas on the east. These include Yugoslavia and its various republics and Romania, Bulgaria, Albania, Greece, and Turkey.)

Event Two: *Georgia* begins to seek independence in late 1990, and formally declares itself independent in April 1991. Soviet troops are used to quell unrest.

Event Three: A coup is attempted by military leaders on August 19, 1991, against USSR President Mikhail S. Gorbachev. Boris N.

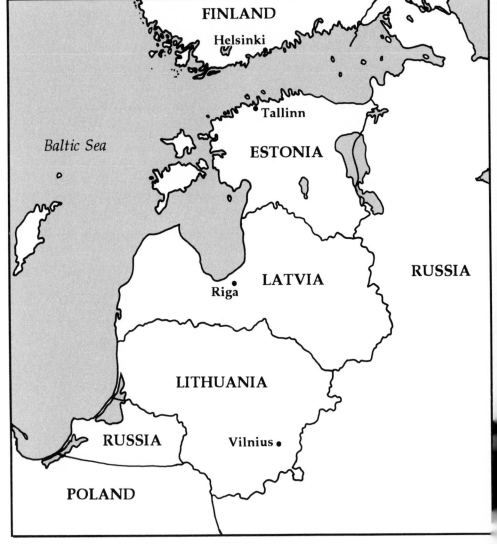

The Baltic States

Yeltsin, president of the republic of *Russia,* mounts a tank in *Moscow* and vows his support for Gorbachev. The coup attempt triggers a rush toward independence in the days and weeks that follow.

Event Four: Lithuania reaffirms its independence immediately, and the other two Baltic states, *Estonia* and *Latvia,* follow in short order.

Event Five: In late August 1991, the following republics of the USSR also declare their independence:

Azerbaijan

Byelorussia (Belarus) (or Byelarus)

Kyrgyzstan

Moldova

Ukraine

Uzbekistan

Event Six: By the end of 1991, these additional republics declare their independence:

Armenia

Kazakhstan

Russia

Tajikstan

Turkmenistan

Event Seven: On December 21, 1991, eleven of the twelve former republics of the USSR—Georgia being the only exception—join the Commonwealth of Independent States.

Event Eight: The USSR is formally dissolved on December 25, 1991.

*   *   *

After you've identified the above countries and regions on the maps available to you in this book—which are comparable to what you might find in a magazine or newspaper—you've completed Steps #1 and #2. Now you're ready to proceed to Steps #3 and #4. This involves a comparison of the identified areas with more detailed maps from your home atlas and also with maps showing a broader expanse of territory. The idea here is to gain increasing amounts of geographical information about the regions under consideration, and also to put the entire area into a wider geographical context.

As you examine each of the regions in more detail, you should note the capitals and other major cities. Also, note any important physical features, such as major rivers, seas, and mountain ranges. You may perceive for the first time how the famous Ural Mountains cut through part of Russia and separate the Moscow area from the Siberian region. If you have a new map, you will see St. Petersburg, which was called Leningrad under the old Soviet government, as well as Kiev, the capital of the Ukraine, and Minsk, the capital of Byelorussia (Belarus). This exercise will help you fix in your mind exactly where each of these new nations is in relation to the others.

When you move on to Step #4 and take a bird's-eye view of the scene with a bigger map, you'll see how the Black Sea and the Caspian Sea are situated on the southwestern borders of the area. And you'll be able to understand better how the new republics are juxtaposed to parts of Europe, China, and other regions.

Step #5 involves plugging facts about current events and history into the physical setting. You already have some of this information in the scenarios described above, and it may help to review it briefly at this point. Here are some further background points that will help you flesh out the geography with a little history.

- A quick look at the name of the country can often tell you about its government as well as its history. *USSR* stands for the *Union of Soviet Socialist Republics.* We have mentioned the various *republics* already; these are simply the names given to certain regions. That they are *soviet*—the Marxist-Leninist term for democratic assemblies—and *socialist* tells you that we're dealing with a nation where the state owns all property.

- The *union* shows that the various republics have joined, or united, under a single government, similar to the way in which the states in America united to form the *United States of America.*
- So simply from the name we know how the nation was formed (a union of republics), the form of government (soviet, or democratically elected from the Marxist-Leninist viewpoint), and their economic system (socialist).
- Although the various republics have existed independently in some form for centuries, they came under the USSR relatively recently. Remember, the Communist revolution occurred in 1917, and so all of the constituent republics joined after that.
- Georgia joined the USSR in 1936, after it was invaded by the Red Army.
- The Baltics—Lithuania, Estonia, and Latvia—became part of the USSR in 1940, after being annexed in a secret 1939 pact between the Soviets and the German Nazi regime.
- The Ukraine, the second largest of the Republics in population (more than fifty-two million people), had been under Russian domination since the seventeenth century. It joined the USSR in 1922.
- Armenia joined in 1922, after Turkey and Russia split Armenia between them.
- Azerbaijan joined in 1922, after the Red Army invaded it.
- Kazakhstan joined the USSR in 1936, though it had been under Russian influence and rule since the eighteenth century.
- Russia, the most populous (about one hundred fifty million people) and influential republic in the USSR, had been ruled by the czars before the revolution of 1917.

From this brief overview, you can see that Russia was the centerpiece of the USSR. Although a few of the other republics had been under Russian domination for centuries, most joined the Soviet state under various forms of duress after 1917. This historical perspective, coupled with a focus on the geography, should prove helpful in understanding how the *union* of various republics, with their very different histories and interests, could fall apart so quickly.

Finally, for Step #6 in your geographical approach to the breakup of the USSR, close your eyes for a few moments and picture the piece-by-piece disintegration of the Soviet empire. Recall as much as you can about the political and historical factors we've discussed that went into this cataclysmic series of events.

You certainly won't remember everything or every republic we've talked about. Nor will you be ready to teach a course in contemporary Russian history. But you should be considerably more knowledgeable about these occurrences than you were before you started. And you'll also be in a stronger position to comprehend the meaning of important future events that will surely occur in that part of the world.

LET'S MOVE ON TO THE MIDDLE EAST

In less detail, I want to challenge you once more to apply our six-step geographical strategy to another part of the world that has been in turmoil and probably will continue to dominate the news at some point in the future. Every knowledgeable person *must* know something about the Middle East, and that means developing a grasp of the geography of the region.

What do you need to know? Begin by identifying these key nations and a few of their characteristics. An emphasis has been placed on the religious and ethnic characteristics of these countries because those factors tend to define many of the conflicts and controversies in the region. To analyze the geographical characteristics, use the six steps already described.

Israel: Israel lies on the far eastern end of the Mediterranean Sea. Of the more than four and a half million people in this tiny but pivotal nation, approximately 82 percent are Jewish, 14 percent

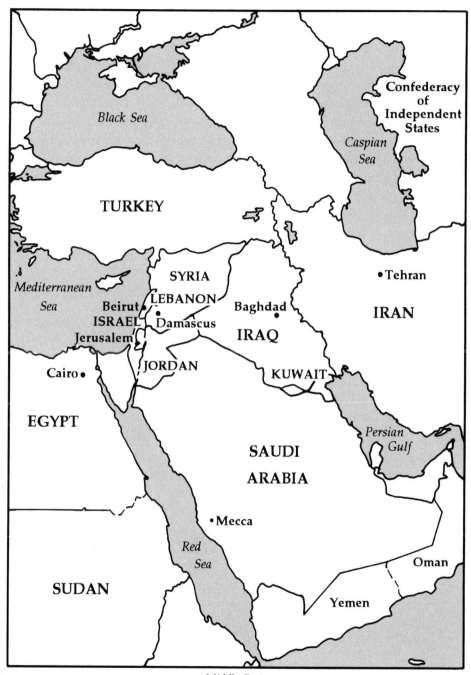

Middle East

Muslim, 2 percent Christian, and 2 percent others. Israel, which has close ties with the United States, has been in a series of wars and conflicts with its Arab neighbors since it became an independent state in 1948. Almost every major confrontation in the Middle East involves Israel, either directly or indirectly.

Egypt: This nation, a country in northeastern Africa, is located just to the southwest of Israel on the Mediterranean Sea. It is home to over fifty-five million people, 90 percent of whom are Arab, the rest, Bedouins and Nubians. About 90 percent are Muslim (mostly Sunni), and the rest are Coptic Christian and of other Christian persuasions.

Jordan: This predominantly Sunni Muslim nation (95 percent of the population) is situated just to the east of Israel, the two nations being partially divided by the Jordan River, the rest by an armistice line. It has more than four million people, over half of whom are Palestinian, and is ruled by a monarchy.

Saudi Arabia: This monarchy—which occupies a vast area to the east of the Red Sea, south of Jordan, and west of the Persian Gulf—includes fifteen million people, nearly all of whom are Arab and Muslim. Saudi Arabia was one of the major allies of the United States and the other nations that fought against Iraq in the Persian Gulf War in 1991.

Kuwait: Kuwait rests on the Persian Gulf, between Iraq and Saudi Arabia. The more than two million people in this oil-rich monarchy were overrun by Iraq in the Gulf War, then liberated by the allies. Nearly all of the people are Muslim, with a predominance of the Sunni sects.

Iran: This Islamic republic, dominated by the often militant, fundamentalist Shiite Muslims, lies across the Persian Gulf to the east of Saudi Arabia. The Arabian Sea is on the south; the Caspian Sea on the north; and Turkmenistan, Afghanistan, and Pakistan to the east. Iran has more than fifty-five million people. Ethnically, the nation is more than 60 percent Persian; 19 percent Turkoman; 4 percent Arab; and 3 percent Kurdish.

Turkey: There are nearly fifty-seven million people in this nation, which is bordered by the Black Sea on the north; the Mediterranean, Syria, and Iraq on the south; the Aegean Sea and Bulgaria on the west; and a bit of Iran on the east. Almost all of the people in Turkey are Muslim, and ethnically the nation is 85 percent Turkish and 12 percent Kurdish.

Syria: The nearly thirteen million people in this country, which is located on the eastern coast of the Mediterranean just above Israel, Lebanon, and Jordan, are 90 percent Muslim (mostly Sunni) and 10 percent Christian. Ethnically, the country is nine-tenths Arab. Syria has long been a leading player in Arab conflicts with Israel.

Iraq: This nation, which opposed the United States and its allies in the Persian Gulf War of 1991, is located just to the east of Syria and Jordan; to the north of Saudi Arabia; and to the west of Iran. A part of Turkey defines the boundary of Iraq on the north. Nearly eighteen million people live in this country, which is 75 percent Arab, 15 percent Kurdish, and about 10 percent Turkish. Religiously, Iraq is 95 percent Muslim—with about 60 percent fundamentalist Shiite and 35 percent Sunni.

Because Iraq, under the dictator Saddam Hussein, has played such an important role in the recent history of the Middle East, more needs to be said about this country.

Turmoil in the area covered by present-day Iraq didn't start with the 1991 Persian Gulf War. Archaeological excavations reveal evidence of settlement dating back to about 10,000 B.C. The greater part of present-day Iraq was originally known as Mesopotamia, from a Greek word that means "land between the rivers." The rivers referred to are the historic Tigris and Euphrates. The point at which they come closest together is the site of the contemporary city of Baghdad, the capital of Iraq. The Tigris and Euphrates are referred to frequently in the Bible and other ancient texts. Many believe that the geographical site for one of the earliest references to these rivers, the Garden of Eden in the book of Genesis, was located somewhere in the vicinity of present-day Basra. This city is in Iraq near the Persian Gulf.

The ancient Babylonian and Assyrian civilizations thrived on the land occupied by Iraq. That great ancient Assyrian city Nineveh became a center of ancient culture in the eighth century B.C., and the political and economic hub of the Mesopotamian region. This cradle of civilization was conquered and reconquered many times by Medes, Persians, and others over the centuries. Yet this region has a way of not only surviving, but thriving and influencing the broader world, as happened with Iraq in the Persian Gulf War. If history is any guide, we can look for further turmoil in this very same area in the not-so-distant future.

Understanding where these nine Middle Eastern nations are located and how the geography relates to the basic ethnic and religious facts that I've summarized can give you a good idea about how conflicts in the region get started and keep going. Some of the factors that contribute to the unrest include these:

- The presence of bitter Palestinians, many of whom were displaced when Israel was established, is a volatile ingredient in the Middle Eastern mix. The fact that there is such a large number of Palestinians in Jordan, which borders on Israel, tends to exacerbate bad feelings.
- There are long-standing enmities between the Arabs—who make up the majority of many of the surrounding nations—and the Jews, who make up the majority of the population in Israel.
- The religious differences, not only between militant fundamentalist Shiites and the Jews but also among different Arab sects, also contribute to conflict.
- There is much to be resolved in this part of the world, and the September 1993 signing of a peace agreement between the Israeli prime minister (Yitzhak Rabin) and the chairman of the Palestine Liberation Organization (Yasir Arafat) is a promising step toward a settlement in this troubled area.

TRY IT AGAIN!

As you read or listen to news reports, use the six-step strategy to analyze the geographical issues. For example, you might take a part of Africa, such as Somalia, which is undergoing problems with famine. Or when you encounter a story about the European Community, note the names of all the member nations (there are twelve as of this writing), and check their location on your maps.

As you do your analysis, study various maps of the region in the way I've suggested, and then integrate the historical facts and current events into the physical setting. I'll guarantee you that this approach will make your reading and listening to important happenings in the world much more pleasurable. You may even start becoming something of an expert on current affairs!

Now, on to an important geographical exploration, which will center on some of the important physical features of the earth.

CHAPTER THREE

GETTING PHYSICAL ABOUT THE EARTH

Throughout most of this book, we will concentrate on the political and scientific divisions and descriptions that human beings have applied to the earth. But there's another, more basic feature of geography—the physical foundations on which all our other knowledge rests. Before exploring the man-made divisions, it's important to understand the physical ones they are often based on.

The oceans, seas, land configurations, and land masses form the basis for our maps, and often *change* the face of the globe as geological events, such as earthquakes and volcano eruptions, occur. To make this part of your geographical education a little easier to grasp, I want to limit this part of our discussion to three separate, but very practical, sets of topics:

- First we'll examine some important physical terms that may become confusing unless you understand the precise definitions and distinctions.
- Then we'll turn our attention to the lifelines of civilization, some of the important rivers.
- Finally, we'll explore perhaps the most important source of geographical change—the body of knowledge known as "plate tectonics."

DO YOU KNOW THE DIFFERENCE?

Many people become confused when they are confronted with these four important pairs of geographical terms: 1) bay and gulf; 2) ocean and sea; 3) cape and peninsula; and 4) isthmus and strait. Here's a quick guide to the distinctions.

Bay Versus Gulf

A *bay* may be described as an indentation of the sea into the land, where the indentation has a wide opening. For example, the Bay of Biscay is an indentation of the North Atlantic between Spain and France.

A *gulf* is a portion of the sea partially enclosed by a sweep of the coast. A gulf tends to be longer and narrower in shape than a bay, though it's not always clearly distinguished from a bay. Look, for instance, at the shape of the Hudson *Bay* and the *Gulf* of Mexico. Or check the *Bay* of Bengal and the *Gulf* of Guinea (off West Africa).

It's not always easy to tell why geographers have chosen one name over the other, and sometimes the answer may be quite mundane: People in a certain part of the world have just become accustomed to calling the body of water near them a "gulf" or "bay" and the name has stuck, regardless of our attempts to make a distinction.

Ocean Versus Sea

Oceans are connected masses of water that cover about 70 percent of the earth's surface. The three oceans recognized by most geographers are the Atlantic, the Pacific, and the Indian. But popular usage also recognizes the Arctic Ocean (within the Arctic Circle) and the Antarctic Ocean (within the Antarctic Circle). Technically, however, these last two "oceans" are considered extensions of the other three.

The term *sea* is used to denote a subdivision of an ocean: For example, the Caribbean Sea, the North Sea, and the Yellow Sea. *Sea* is also used to name very large lakes (usually those with a high concentration of salt). These include the Caspian Sea, the Aral Sea, and the Dead Sea. We also have seas that are almost surrounded by land, such as the Mediterranean Sea (with an outlet to the North Atlantic), the Adriatic Sea (with an outlet to the Mediterranean Sea), and the Red Sea (with an outlet to the Gulf of Aden).

Cape Versus Peninsula

A *cape* is a piece of land, often hooked or pointed, which juts into the sea. It's also described variously as a headland and a promontory. For example, Cape Cod, Cape Hattaras.

A *peninsula* is a piece of land that is almost an island since it's nearly surrounded (on three sides) by water. Hence, the term refers to any piece of land that projects into the sea so that its boundary is mainly coastline.

Here again, distinguishing between a cape and a peninsula is principally a matter of size. Peninsulas tend to be bigger, capes are smaller. Florida is a peninsula and so is lower California (Baja California). Similarly, Italy is a peninsula, as is the Iberian Peninsula (Spain and Portugal), and the Scandinavian Peninsula (Norway plus Sweden). Technically, India is also a peninsula, though it's also categorized as a subcontinent of Asia.

Isthmus Versus Strait

An *isthmus* is a narrow strip of land that connects two large land masses. This land bridge often has great strategic and commercial importance because it may be the only land route between two large land areas. In modern times, many isthmuses have been cut through by canals, thus eliminating very long land or sea routes. For instance, the Panama Canal on the Isthmus of Panama, which connects Central and South America, makes it unnecessary for ships from New York to sail around South America to reach San Francisco. And the Suez Canal on the Isthmus of Suez joins Asia and Africa. This canal, which links the Mediterranean Sea to the Red Sea, was cut between 1859 and 1869. A look at your world map will immediately show the importance of this engineering feat.

A *strait* is, in a sense, the opposite of an isthmus. It's a narrow waterway connecting two large bodies of water. In some cases, a strait may function as a kind of canal, but a true strait is a natural geographic feature, while a canal is man-made. Some famous straits are the Strait of Gibraltar and the Bering Strait.

THE ROMANCE OF THE RIVERS

Rivers have always fascinated mankind. They have inspired romance, fear, awe, respect, love, and many other emotions. We use them for our pleasures and also for business and commerce. We also abuse them by expecting them to carry away the poisonous wastes of our industrial societies.

Because rivers are essential to life, it's not surprising that civilization began between two great rivers, the Tigris and Euphrates, in an area known as Sumer in southern Mesopotamia. This site is located in present-day Iraq.

Here are at least twenty supremely important rivers on Earth which you should be able to locate on your world map. I've sometimes mentioned only one key country for each, though in many cases, these rivers flow through several nations.

1. *Nile*, Egypt
2. *Amazon*, Brazil
3. *Mississippi*, United States
4. *Congo*, Zaire
5. *Euphrates*, Iraq
6. *St. Lawrence*, Canada
7. *Yangtze*, China
8. *Hudson*, United States (New York State)
9. *Danube*, Hungary
10. *Indus*, Pakistan and India
11. *Niger*, Guinea
12. *Seine*, France
13. *Ganges*, India
14. *Missouri*, United States
15. *Thames*, England

16. *Brahmaputra*, India and China
17. *Tigris*, Iraq
18. *Volga*, Russia
19. *Mekong*, Vietnam, Cambodia, Laos, and China

Of course, there are many other important rivers in the world—and they can truly be lifelines to the communities and nations they serve. To give you an idea of just how important a river can be, consider these capsule descriptions of the first three in the list above.

The Nile River: The world's longest river rises near Lake Victoria in Africa and flows north for forty-two hundred miles to the Mediterranean Sea. Ancient Egyptians called the river Ar, which means black, a description that refers to the color of the sediment when the water overflows its banks. As in ancient times, Egypt today would be entirely a desert were it not for the Nile. In fact, Egypt is often referred to as "the gift of the Nile." As far back as 4000 B.C. a system of irrigation, which is still used today, was introduced to harness and use the flood waters.

The Amazon River: The source of the Amazon is in Peru. When you include the river's longest tributary, it flows four thousand miles across Brazil to the Atlantic Ocean. The Amazon is the world's second longest river and contains more water than any other. It was named Amazuna, meaning big wave, by an Indian tribe that lived on its banks. But the Spanish explorers in the sixteenth century thought its name was derived from the Amazons, or female warriors of Greek legend. The Spaniards reportedly believed the long-haired Indian men they were fighting were women!

The Mississippi River: We often like to think that the Mississippi is the longest river in the world, and it is indeed *if* you include the Missouri (a major tributary) that feeds into the Mississippi. The Mississippi is 2,350 miles long, and the Missouri is 2,464. The total of 4,814 would make it longer than the Nile. (But I personally prefer to regard the two rivers as separate because they have their own separate histories.) The name *Mississippi* comes from an Algonquin Indian word meaning great river. The familiar name is "Ol' Man River," as in the popular song.

The upper Mississippi flows from its source in northern Minnesota as a fairly narrow, clear stream, until it's joined by the muddy colored Missouri. At that point, the lower Mississippi begins its progress through a wide, fertile valley region. Over the millennia, this river has carried down vast quantities of sediment to form a delta that extends out into the Gulf of Mexico.

A JOURNEY TO THE CENTER OF THE EARTH

Actually, at this time I don't want to take you quite to the center of the earth—only about sixty miles down from the surface. This upper part of the earth's shell is called the lithosphere. The uppermost part of the lithosphere is a three- to twenty-two-mile covering known as the earth's crust.

In the lithosphere below the crust, things are definitely not quiet and peaceful. Some major movements occur among huge slabs of rock known as plates. These may be many miles thick, and they respond to incredible temperatures and pressures even deeper inside the earth. The internal temperature of our planet is very high: The core or center of the earth reaches 4,300 degrees centigrade, and some researchers say that the heat within the earth's core is even hotter than the surface of the sun! This great internal heat is responsible for the movement and instability in the upper levels of the earth, including the shell or crust where we live.

There are thought to be seven large plates in the lithosphere that cause most of the movement on the surface, including earthquakes and volcanoes: the North American Plate, the South American Plate, the Pacific Plate, the Indo-Australian Plate, the African Plate, the Eurasian Plate, and the Antarctic Plate. In addition, there are several smaller plates that participate in the action.

These plates move so gradually that we usually can't feel them. Geologists tell us that the plates move continuously against one another at rates of up to twenty centimeters per year. As the great slabs of rock, or plates, interact below the surface, they produce an action called plate tectonics. The word *tectonics* comes from a Greek term referring to building something, and that's just what the plates do. They build mountains, reshape the land, fuel volcanoes, and cause earthquakes. The plate movements on all sides of the Pacific

Ocean plate are the cause of the earthquakes and volcanic eruptions along what I've already referred to (in Chapter One) as the Pacific Ring of Fire. Also, plate tectonics have caused the Himalayan Mountains in Asia to grow higher each year. North American and Pacific tectonic plates slide along against each other at the San Andreas Fault, thus creating the strong possibility of major earthquakes in California.

Tectonic plates are also responsible for a phenomenon known as continental drift. A German scientist, Alfred Wegener, suggested that all the continents were once joined and that over millions of years, they split apart, with the plates moving the various land masses to their present positions. There's much evidence to support this theory. For example, Africa looks as though it would fit in like a piece of a jigsaw puzzle when it's placed next to North and South America. Furthermore, fossils of at least one ancient land animal have been found in South Africa, India, and Antarctica. Because this animal was from a warm climate, scientists say that Antarctica was once joined to Africa and India. In addition, mountains and rock formations found in North America have been found to match similar belts of mountain and rock discovered in Europe on the other side of the Atlantic.

And so the adventure of geography continues through seemingly endless variations and scientific disciplines. It's hard to understand what's happening on the surface of the earth without knowing what's going on below. But we also know that the movement of the plates is only one factor that enters into our geographical understanding. After all, the surface is constantly changing independently of plate tectonics or other physical influences because of the *human* impact on the map and the environment. We unceasingly alter our political boundaries, build new structures, dig new canals or wells, pollute and destroy the environment, and otherwise have a cultural impact on our surroundings.

Now, let's turn from this discussion of the physical foundations of geography to some "quick hits"—short geographical facts that might be classified as trivial if they weren't so *un*trivial.

CHAPTER FOUR

NOT SO TRIVIAL TRIVIA

I find that many times, having at hand a specific piece of geographical information—a bit of trivia, some might say—can make a significant impression on other people. Also, one fact or short set of facts can make complex national or international news events or other situations considerably more understandable. The key question, of course, is to identify *which* fact is worth knowing; in other words, which bit of trivia is not so trivial. The purpose of this chapter is to suggest a few types of not-so-trivial trivia and also to provide you with some guidelines for picking geographical facts that will help make you shine in your work and social interactions. To illustrate what I mean, let's try a case study involving a particular, complex set of news events.

After the most recent drive for independence began among the Yugoslavian republics in 1991–1992, the nature of the sides involved in the fighting thoroughly confused practically everyone I encountered. For that matter, I was quite confused—until I sat down and sorted through the geographical underpinnings of the conflict.

I started with some simple maps found in almost any newspaper and with minimal salient facts taken from any good periodical. Then I opened my atlas and identified just a few of what seemed to be the "untrivial trivia," or important items of information. In a matter of minutes, the veil of mystery lifted, and the situation in that war-torn land became much more understandable. Here are the key bits of nontrivia that I used as a foundation to increase my understanding of the Yugoslavian situation. Note locations on the accompanying map.

Nontrivia #1: There were four breakaway republics, which had declared their independence from Yugoslavia: (1) Croatia (declared independence on June 25, 1991); (2) Slovenia (June 25, 1991); (3) Macedonia (November 17, 1991); and (4) Bosnia and Herzegovina (March 3, 1992). These republics rely on historic, pre–World War I borders to legitimize their existence.

Nontrivia #2: The "new Yugoslavia," which remained after the breakup, consisted of two main republics: Serbia and Montenegro, which rely on historic factors, but more on ethnic composition to define their borders.

Nontrivia #3: A number of important ethnic and religious groups, including Serbs, Muslims, and Croats, were spread across the lines of the various republics. Part of the stated motivation for the conflict was to secure areas occupied by these groups (and especially by the Serbs), regardless of the republican boundaries.

Nontrivia #4: I formulated a capsule version of Yugoslavian history in the twentieth century: The Kingdom of the Serbs, Croats, and Slovenes was created after World War I, in 1918. This entity lasted until World War II, when the Nazis invaded and the Partisans, led by Tito, offered opposition. The independent communist state of Yugoslavia was established in 1946 under Tito and lasted until the rise of the "new Yugoslavia," mentioned in *Nontrivia #2.*

This "new Yugoslavia" claims that the old communist Yugoslavia united the nation across both ethnic and historical borders, and seeks to regain the "natural" ethnic boundaries of the pre–World War II kingdoms. The smaller breakaway republics, meanwhile, claim both the communists and the post–World War I kingdoms violated their ancient sovereignty, and they seek to regain their historical pre–World War I boundaries. This is complicated, as all groups have a historical claim depending upon how far back one goes. Ethnic claims are also complicated, as ethnic groups migrated from their historical homes during the period of the unified Yugoslavia.

Obviously, a great deal more could be said about the history and

Former Yugoslavia

geography of Yugoslavia, especially with respect to the location and interactions of the various ethnic and religious groups. Also, the situation is constantly changing, and by the time you read this, the scene in that part of the world could be quite different. But chances are, a knowledge of just the above key facts will continue to be helpful well into the foreseeable future for those who want to discuss and read about the Yugoslavian situation with more authority and understanding.

Now, what other bits of "untrivial trivia" can help you establish yourself as a person who is knowledgeable about the earth and current events? If you have a grasp of the following geographical points, you'll have a better comprehension of the physical make-up of our environment. And you'll be way ahead of most of your friends and colleagues in understanding geography.

THE MOST IMPORTANT PIECES OF UNTRIVIAL TRIVIA

The following individual facts have struck me as likely to be most useful when I'm trying to understand the geography of a situation, or when I'm engaging in general conversations that may have some bearing on geography. At first glance, some of these items may not seem particularly useful to you—such as the deepest well or the highest inhabited town. You may never use many of these in casual conversation, even in a trivia-oriented cocktail party exchange. In other words, they strike you as classic examples of truly *trivial* trivia. But I find that *all* these facts, when taken together, provide me with a general geographical perspective and framework. Just knowing what's *possible* in physical terms affords a better perspective on the dimensions and potential of this earth where we live. Undoubtedly, you'll think of other facts that apply more directly to your personal or business situation. Be sure to jot them down on a separate sheet of paper as they come to mind. You may want to insert that sheet into the book at this point so that you'll in effect have an addendum to the basic list of untrivial trivia. The final section of this chapter includes some guidelines for formulating your own list. But first, here's mine:

1. *The European Community:* Formerly known as the European Economic Community, the current member countries in this European economic and political organization include the following twelve nations:

United Kingdom (UK) and Northern Ireland

Republic of Ireland

Spain

France

Belgium

Luxembourg

Germany

Netherlands

Denmark

Italy

Greece

Portugal

Anyone who travels or does business in Europe should be aware of these member nations. *Note:* Four other European countries—Switzerland, Austria, Norway, and Sweden—are currently *not* members of the European Community. But they do belong to the European Free Trade Association (EFTA). Later, any one of these might apply for membership in the European Community.

2. *Size of Earth:* Total land surface area of the earth is 57.5 million square miles, or 29 percent of the surface; total water surface area is 139.4 million square miles, or 71 percent of the earth's surface.

3. *Tides:* The highest tides have been recorded at seventy feet in the Bay of Fundy, Nova Scotia.

4. *Deepest lake:* The deepest lake in the world is Lake Baikal in Siberia, which is 5,715 feet deep (more than a mile!).

5. *Largest lake:* The largest *freshwater* lake in the world is Lake Superior, one of the five Great Lakes of the United States and Canada. It is 32,000 square miles in area. (There is one larger inland body of water, the Caspian Sea, which lies to the north of Iran and to the south of Russia. But as a salt lake with an area of 152,239 square miles, it's regarded as a "sea" rather than a lake.) By the way, the best way we know to remember the names of the five Great Lakes is to use the initials H O M E S thus: *H*uron *O*ntario *M*ichigan *E*rie *S*uperior.

6. *Longest river:* The longest river in the world is the River Nile in Egypt. It's 4,200 miles long—and continues to provide important irrigation and water sources for millions of people along its banks.

7. *Age of the earth:* The age of the earth is estimated to be between 4.5 and 4.7 billion years.

8. *Longevity of the Sun:* It's estimated that the sun will burn out in about five billion years.

9. *Melting of polar caps:* If the earth's polar ice caps were to melt (say, through extreme global warming), the mean sea level would rise by about sixty miles, a catastrophe that would submerge half the world's population.

10. *How deep is the sea?* The average depth of the ocean is more than thirty-six hundred feet, or five times the average height of the land above sea level. The greatest known depth in the Pacific Ocean, Mariana Trench, is 36,198 feet.

11. *Speed of the earth:* The earth revolves in its orbit around the sun at a speed of 66,700 miles per hour. It makes one complete revolution around the sun every 365 days, 5 hours, 48 minutes, and 46 seconds. (Hence, there is a need to make a calendar adjustment every four years by including an extra day in February. This makes a "leap year" of 366 days for that fourth year.)

12. *Earth's rotation:* The earth rotates on its axis at a speed of more than one thousand miles per hour at the equator. It makes one complete rotation on its axis in 23 hours, 56 minutes, and 4 seconds. (The leap-year time adjustment every four years is also helpful here to enable us to maintain our twenty-four-hour day.)

13. *How fat is the earth?* The circumference of the earth at the equator is 24,901.45 miles.

14. How heavy? The weight of the earth is 6,600 billion billion (no, that's not a misprint!) tons.

15. How high? The highest point on the earth's surface is Mount Everest. This mountain is 29,028 feet in altitude and is located in both Tibet and Nepal.

16. How low? The lowest point on the earth's land surface is at the Dead Sea, the salt water sea between Israel and Jordan. It's 1,312 feet below sea level.

17. How hot? The hottest temperature ever recorded was 136 degrees Fahrenheit in Libya on September 13, 1922.

18. How cold? The lowest temperature ever recorded was –129 degrees Fahrenheit at Vostok, Antarctica, on July 21, 1983.

19. How many? The latest available estimate, mid-1992, for the total population is 5.420 billion people—and it's going up every day.

The nation with the largest population in the world is the People's Republic of China. The last estimate tabbed China at more than 1.13 billion.

20. Largest country: The largest country in land area in the world is Russia, which measures 6,592,849 square miles. (The old USSR was 8.6 million square miles. In contrast, the United States is only 3,679,245 square miles.)

21. The deepest well: An oil well in Pecos, Texas, is thought to be the deepest at 25,340 feet.

22. The highest town: The altitude of Aucanquilcha, Chile, is 17,500 feet above sea level.

23. The wettest spot: On average, the wettest spot on the earth is Mount Waialeale, Hawaii, with an annual average rainfall of 471 inches.

24. New countries: By the end of the 1980s, 118 new countries had been established throughout the world during the twentieth century. A wave of new nations have also been established in the 1990s, beginning with Namibia in southwest Africa. The move toward independence has been accelerating with the fifteen new republics from the USSR, and four independent states out of old Yugoslavia.

25. Poorest countries: There are many ways to measure poverty, such as average per capita income or gross national product. I prefer to focus on the most obvious and disturbing—death and devastation to health through war, famine, flood, or other natural disaster.

As I write, the nation that seems to be suffering the most is Somalia, located on the east coast of Africa, just to the east of Ethiopia, west of the Indian Ocean, and South of the Gulf of Aden. By rough International Red Cross estimates, more than one hundred thousand people have died there of ongoing famine, with hundreds (some say thousands) continuing to die of starvation or malnutrition-related diseases every day.

When you read this, another nation may be suffering more than Somalia. But I'd encourage you not to ignore this final item just because it seems out of date. Rather, *substitute* the latest information, or the part of the world where suffering concerns you the most. And finally, ask yourself what you can do in the way of contributions, lobbying, or volunteer work to relieve the anguish of those who live there.

HOW TO MAKE YOUR OWN PERSONAL NOT-SO-TRIVIAL TRIVIA LIST

I find it's helpful to keep my own personal nontrivial trivia list so that I'll be sure to use my understanding of geography in the most practical way. Generally, I try to answer these questions as I'm compiling the list:

Question One: What three geographical facts or sets of facts in my business or at work should I have at my fingertips?
It may be that you live in Chicago but you travel regularly to Dallas. If so, you should try to get to know Dallas as well or better than you know Chicago. Too often, business travelers don't take time to learn as much as possible about the areas they are visiting, including the most important historical and cultural sites. As you explore the geography of an area, you'll automatically pick up facts about the social, economic, and political life—infor-

mation that will be invaluable in helping you relate better to colleagues and clients.

Question Two: What geographical facts would help me relate better to friends or colleagues with whom I'm regularly in contact?

If one of your close associates is from San Francisco, another from Miami, and a third from New York, pull out your atlas and encyclopedia, and learn a little more about those places. You'll find that important areas of conversation will open up as you get people to talk about their hometowns.

Question Three: What parts of the nation or the world do I want to know more about?

Most of us are fascinated by some far-flung place, but we just haven't found the time to read or study about it. Take some time during the next week to consult your atlas, encyclopedia, and perhaps an almanac and do some quick reading on the subject. Then, if you really feel ambitious, check out some books on the area from your library and do further reading—with your atlas near at hand.

It doesn't take long to draw up your own personal list of nontrivial trivia. And if you continue this exercise for a few minutes at a time for several months, you'll find yourself developing a broad and impressive grasp of geography, history, politics, and other subjects. When trivia accumulates over an extended period of time, disconnected and seemingly unimportant facts have a way of being transformed into broad knowledge, incisive insight, and even wisdom.

CHAPTER FIVE

MAP READING FOR THE LOST

One of the most common and practical uses of geography is the art of reading and following a map. We have to employ maps for a variety of purposes:

car journeys over unfamiliar territory

walking trips through unknown cities or towns

hikes across terrain that contains few roads

understanding news stories or other accounts that depend heavily on accompanying maps

discussing unfamiliar features of cities, states, or nations— features that can only be comprehended through geography

Yet despite the fact that we are forced to use maps regularly, many of us lack the skills to read and follow them. I'm reminded of one friend who was being driven along some winding roads in New England. She had been designated as the "navigator" for the trip, and so all the road maps and directions were dumped in her lap. The particular route that the group had chosen was rather complex and circuitous, with many abrupt turns onto poorly marked side roads. Before my friend knew what had happened, she was lost. Then, the questions started coming fast and furiously from her colleagues:

"Did we already pass state road twenty-one?"

"How long have we been on this highway?"

"Did you see the interstate intersect our route?"

"Why didn't you just follow the handwritten directions?"

The experience was embarrassing and frustrating, and more than an hour was lost before the group finally found its way back onto the right road. Mercifully, my friend was allowed to continue as navigator and thus save some face. But she *didn't* volunteer to navigate on the return journey, and she vowed that on future trips, she would take *any* job other than map reader!

Most people I know have had an experience like that at one time or another. Yet the mistakes and frustration could all be avoided by acquiring a few more map-reading skills, and by taking more care in using that knowledge. To understand better how to read a map, it's necessary for you to have some background on the purposes of maps, how they are produced, and how accurate they are.

CAN YOU REALLY RELY ON A MAP?

The map is our best vehicle to show where places are located throughout the world. As a dictionary defines words, so an atlas defines the physical features of the earth. Only on a map can we see and understand where countries, cities, rivers, mountains, and lakes are in relation to one another.

The accuracy of a map depends in part on the size of the slice of the earth that is pictured. In general, the smaller the area depicted, the more accurate the map is. The reason for this is that a relatively limited area minimizes the distortion created by the earth's curved surface. On the other hand, a very broad area, such as a continent or hemisphere, will have more distortions because the earth's curve will be more pronounced.

Globes and map projections: Of all large-scale maps, only a *globe* can represent the earth's surface features correctly with reference to area, shape, scale, or direction. A projection from a globe to a flat map always causes distortions of some of these characteristics.

The mathematical system by which the spherical surface of the earth is transferred to the plane surface of a map is called *map projection.* One of the best-known projections was invented by Flemish geographer Gerardus Mercator in 1569. His work has been widely used ever since for maps that project the entire world on one plane.

In this projection, all parallels (lines of latitude, which encircle the globe, parallel to the equator) and meridians (lines of longitude, which encircle the globe by intersecting at the poles) are represented as straight lines intersecting at right angles. On the Mercator projection, all the parallels have the same length as the equator, thus causing considerable distortion in high latitudes.

Compare, for instance, the size of Greenland with that of South America or Australia on Mercator's projection. Polar regions can never be shown because this would produce a "square" earth. You might also note the distortion of the continent of Antarctica on the Mercator projection.

The quest for accuracy: Many efforts have been made to reduce the distortions of map projections. There's no perfect solution to drawing the round earth on a flat sheet of paper. But some cartographers (mapmakers) have come close. One of the best attempts has been made by the cartographer Arthur H. Robinson of the National Geographic Society. (See page 10.)

Robinson was given the task of finding a substitute for the so-called Van der Grinten projection, which the Society had used since 1922. In this earlier depiction, there were significant distortions: Greenland was 554 percent larger than it was supposed to be; the United States was 68 percent larger; and Africa was 8 percent larger.

In contrast, the Robinson projection, which was published in December 1988, has significantly reduced many of the distortions: Greenland is now only 60 percent larger than it should be, and the United States is 3 percent smaller than its actual size on the globe. But there are some compromises: Africa is now 15 percent smaller than it should be.

To achieve his results, Robinson, professor emeritus of cartography and geography at the University of Wisconsin, Madison, took an artistic approach. He began by concentrating on making the shapes of the largest land masses as realistic looking as possible. Then he applied mathematical models to make the size of the various areas as accurate as possible. The most accurate depictions on his projections are at the parallel that is 38 degrees north of the equator, and at the parallel that is 38 degrees south of the equator. Those parallels cut through the parts of the earth that contain most of the land and people.

Which way is north? Since the mid-fifteenth century, European mapmakers have generally arranged their maps with north at the top of the sheet. Earlier maps weren't standardized this way. The circular world maps of the Middle Ages were oriented with east at the top because it was in the East that the Garden of Eden was traditionally situated. Indeed, the word *orientation* originally meant the arrangement of something so as to face east. This is also the reason that countries east of the Mediterranean may be referred to as the *Orient.*

When the mapmaker makes a mistake: Even though a small area can *theoretically* be pictured more accurately on a map, sometimes the cartographer makes a mistake—or intentionally inserts an error. (One of my sources has informed me that by creating an error in a little-traveled part of the map, cartographers can make their maps quite distinct from other maps and thus preserve their copyright.)

A man I know encountered one of these mistakes on the backroads of West Virginia. He and his brother were traveling at dusk, and they decided they had better try to find a motel. But when they took two turns indicated on the map, they ended up on a road that had a route number—but wasn't on the map! Furthermore, the intersections indicated on the map didn't correlate with any actual intersections. After driving around in circles for an hour, they finally found their way back to a major thoroughfare and found a place to spend the night. But both were exhausted—and quite angry at the mapmaker who had led them astray. But even the most accurate map won't help you if you don't know how to read it.

HOW TO READ A MAP

Before you even try to read a map, it's important to identify the type of map you're dealing with. There are four main categories of maps that you may encounter: 1) road maps; 2) contour maps;

3) physical relief maps; and 4) political maps. Each of these requires a distinct type of approach for the map reader.

Road Maps

Any map, including a road map, is a collection of symbols that stands for reduced representations of something on the globe or a part of the globe. Map symbols and markings may take many forms, which seldom look like the feature they represent. As a result, maps must always be interpreted by using a *legend* or *key*, which is located in one corner of the map and identifies the various symbols and markings.

The symbols on a road map may be dots or circles (representing towns, or sometimes cities); irregular shapes (representing large cities); or colors (representing special sections of states, forests, parks, or deserts). A single line will indicate a secondary road or street, while a double line will symbolize a major highway. Numbers that lie on a road between two indicators (such as diamond-shaped symbols) show the number of miles between the indicators.

It goes without saying that you can't find the lines, dots, or other symbols on the surface of the earth—except in a few rare cases. In some parts of the world, if you're standing on the equator at 0 degrees latitude, you may actually look down and see a line! Some tourist sites on the equator have actually drawn that symbol on the earth so that visitors can stand simultaneously with one foot in the northern hemisphere and one in the southern hemisphere. There are also other such places, like the spot where the states of Utah, Colorado, New Mexico, and Arizona converge. At this "four corners" location, a visitor can get down on all fours and actually be in four states at the same time!

But these places, where map symbols are transferred to the actual land, are rare. It's usually up to the map reader to study the map first, and then in the mind's eye transfer the map symbols to the actual terrain. Making this transfer is the point at which many people get into trouble—and get lost. To become an effective road map reader, you may find it helpful to keep the following principles in mind. Take a look at your own road atlas to help you visualize some of these points.

- Check the *north-south-east-west directions* on the map, and determine the direction in which you're traveling. Then, turn the map so that *north* on the map coincides with north on the ground. This way, you'll be in a better position to visualize and trace your progress on the map. (Otherwise, you may find yourself going in one direction on the ground, and in another according to the map.)

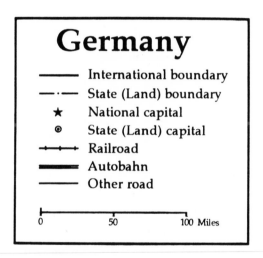

- Spend at least a minute studying the *legend.* This way, the meaning of the various symbols will become more familiar and fixed in your mind.
- Check the *scale* of the map. Is it an inch scale (e.g., ten miles = one inch), or is it a graphic scale (with a line graph indicating the distances between certain marks or checks)?
- *Map out your route,* if you haven't already done so. The usual objective is to find the shortest distance between your starting point and the final destination. But you may decide to travel a few miles more so as to avoid a heavily congested area or

an area under construction. To find the distances on the map, you can take one of two approaches: You can use the distance scale to measure the route, or you can add up the miles indicated between markers on the roads over which you'll travel.

The first method is usually faster, but it also tends to be less accurate. The second method can be quite precise, but can also be very slow and tedious, especially if you're traveling a long distance.

A suggestion: To keep from marking up the map, you may want to list each road, turn, and other direction in a column on a separate sheet of paper.

Contour Maps

As we all know, the surface of the earth is not smooth—yet most flat maps and globes are. To achieve greater accuracy in depicting the elevations and depressions on the actual ground (and in the oceans and seas), a contour map may be used.

This type of map is still flat, but it shows the physical configuration of an area by having a series of "contour lines" drawn on a map through points of equal elevation above sea level. See the accompanying illustration of an imaginary island.

There is always a set distance between the contour lines, which is called a "contour interval." In the accompanying map, the contour interval is one hundred feet.

For example, the line on which A is located joins all points that are one hundred feet above sea level. The vertical distance between any two lines next to each other is one hundred feet. You will see that in some places the lines are close together which indicates that the slope of the land is very steep. But when the lines are set farther apart, the land is flatter.

The contour interval between lines varies with the scale of the map and the type of surface being represented. On a topographic map of a mountainous area, valleys and peaks are as easily recognizable as they would be on an aerial photograph, as the space between the lines indicates the *slope* of the land more accurately than a photograph. (A topographic map, by the way, is one that shows elevations and physical configurations such as contours.)

It may take a little while to visualize what the contour lines mean in terms of flat or hilly terrain, but the effort is worth it, especially if you do any hiking or cycling.

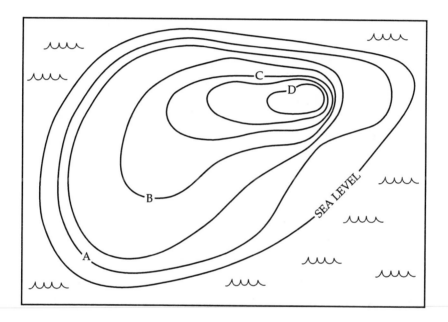

Contour Map of an Imaginary Island

Relief Maps

Some maps, called relief or physical maps, show elevations by special colors instead of contour lines. For example, light green may be used for low-lying areas, and other colors, ranging from yellow to brown, will indicate rising altitudes. Dark brown or white may show the highest points, such as the peaks of mountains. A key or legend should always be provided to help you understand the meaning of the different colors.

Sometimes, in museums or on a special globe or other surface, the relief features may be shown by building actual miniature mountains, valleys, and other features. But in the flat maps found in the usual atlas, you'll have to rely on the legend and your imagination to "see" the physical characteristics of the land.

Political Maps

Like most other maps, political maps show the distribution of land and sea; the shape, position, and size of continents; and the location of many natural features, such as mountains, rivers, and lakes. But the *divisions* of the land are *man-made* and include such distinctive features as boundary lines between countries or states.

Political maps come under the general heading of "human geography," which places people in their physical setting and studies their relationship with their environment. This human branch of geography may be subdivided into political, economic, and social geography, all of which may overlap.

A political map of Europe will show the boundaries between countries, with lines separating the nations, each of which is depicted in a different color. If you look at the *actual* boundary between Spain and Portugal, for instance, we all know that you won't find a line drawn on the ground. But the boundary is known to all who live there—and is represented on any good map.

A map showing the economic geography of a region may indicate the industrial centers, agricultural characteristics, and other features relating to the economy of the area. Political boundaries may also be shown, but the main focus is the economy. A social geography map will show the demographics of the land, such as the distribution of ethnic or religious groups.

The human geography of an area is always changing. In a matter of a few months at the end of 1991, all the old political maps showing the Soviet Union were rendered out of date as a result of the movement for independence among the various Soviet republics. So it's essential to check the *date* of political, economic, or social maps before you begin to rely on them. Otherwise, you may find yourself referring to a country by a name that doesn't exist any longer!

A personal confession: I found myself referring to *West* Germany as a member of the European Community the other day— despite the fact that East and West Germany have been united for several years now. It's still sometimes hard for me to think in terms of one German nation rather than two. The configuration of world geography has been changing so rapidly in recent years that I sometimes find it necessary to cut out maps published in newspapers and magazines and insert them inside my atlas. That's the only way I can be sure that my reference materials are completely current.

How can you obtain the very latest and best maps available? Most large cities have stores that specialize in maps, but if you have trouble finding a good source, you may want to check the following organizations.

WHERE TO FIND A GOOD MAP

Here are some organizations and stores that should be able to supply you with information about up-to-date maps:

The International Map Dealers
   Association
P.O. Box 1789
Kankakee, Illinois 60901
Phone: (815) 939-4627

The National Geographic Society
1145 Seventeenth Street, N.W.
Washington, D.C. 20036
Phone: (800) 638-4077

Rand McNally Map and Travel
   Center
150 East Fifty-second Street
New York, New York 10022
Phone: (212) 758-7488

National Technical Information
   Service
U.S. Department of Commerce
5285 Port Royal Road
Springfield, Virginia 22161
Phone: (703) 487-4650

Government Printing Office
Superintendent of Documents
Washington, D.C. 20402
Phone: (202) 783-3238

Central Intelligence Agency
Public Affairs Office
Washington, D.C. 20505
Phone: (703) 351-2053

Hagstrom Map and Travel
  Center
57 West Forty-third Street
New York, New York 10036
Phone: (212) 398-1222

The Complete Traveller
  Bookstore
199 Madison Avenue
New York, New York 10016
Phone: (212) 685-9007

Map Man
120 Bethpage Road
Hicksville, New York 11801
Phone: (516) 931-8404

Now that you have a better idea about how to evaluate and use the main types of geographical representations, it's time to move from the minutiae of map reading to what I call the Big Global Picture Show—which features continents, hemispheres, time zones, and other earth-size concepts.

CHAPTER SIX

THE BIG GLOBAL PICTURE SHOW–STARRING HEMISPHERES, TIME ZONES, AND SEASONS

Every traveler, especially those who traverse long distances, should have a grasp of the big picture of geography. By this, I mean you should develop a broad mental picture of such facts as how many miles you must go; how many time zones you must cross; and the different cultures, nations, and continents you can expect to encounter. Such a global perspective will not only give you a better intellectual sense of the geographical range of your trip, but will also help you prepare better for the journey.

You'll be more inclined to:

gather helpful street and rapid-transit maps of cities you'll visit

assemble relevant guidebooks

list possible personal contacts who can make your visit more productive or enjoyable

brush up on conversational vocabulary of pertinent foreign languages

focus on precisely what you need to pack—seasons and climates can change greatly as you move from one hemisphere to another

61

explore ways to adjust your sleep pattern

consider steps to reduce jet lag

To gain a global perspective, you should first fix in your mind the way the world is divided, according to hemispheres and lines of longitude and latitude. Then it's helpful to move on to a consideration of time zones and seasons. With this background, you'll be ready to consider more intelligently specific continents and countries.

SAVOR A SLICE OF THE EARTH!

Our planet has been divided into different segments by lines that run from the North to South Poles (lines of longitude, or meridians) and that run *around* the globe, parallel to the equator (lines of latitude, or parallels). A major reason for these divisions has been to enable travelers, navigators, mapmakers (and geography students and buffs) to know exactly where different locations are in relation to one another. This way, distances and directions can be calculated and communicated more precisely.

What's in a Hemisphere?

The first step in getting the big global picture is to divide the earth into two equal parts by the imaginary line running around the earth which we call the *equator*. The equator runs between the north and south polar axis along 0 degrees latitude. The northern half of the world, or everything above the equator, is known as the Northern Hemisphere. The southern half, or everything below the equator, is the Southern Hemisphere. The word *hemi* means half; the word *sphere* refers to any globe or round object. Hence, *hemisphere* means half a globe.

The entire slice of any globe that you view defines a circle. As you know from your elementary-school math, there are 360 degrees in the circumference or outside perimeter of a circle. So one half of the globe's sideways circle, from one of the poles to the other pole, is 180 degrees; and one fourth of the globe's sideways circle, from the equator to either pole, is 90 degrees.

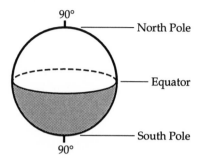

Degrees as a Measurement on a Globe

It's important to keep this simple math in mind, because you'll always know that you can have only 90 degrees, or 90 lines of latitude, from the equator to either pole. If the latitude line, or parallel, is above the equator, you designate it as a *north* latitude; and if it's south of the equator, you designate is as a *south* latitude. So, a line of latitude that runs through the middle of the United States is the 40th parallel north, or 40 degrees north latitude.

The distance between one degree and the next degree (or between one line of latitude and the next) on the earth's north-to-south circle is approximately sixty-nine miles. So if you move from 0 degree latitude at the equator just slightly to the north to the first parallel (one degree of latitude), you'll travel about sixty-nine miles.

But this only takes care of two of our hemispheres, the Northern and Southern. There are two other hemispheres, the Western and Eastern, which are defined by slicing the globe the other way— through each pole.

What About the Other Hemispheres?

Picking the equator as the dividing line between the Northern and Southern hemispheres is a fairly obvious and easy task. But where do cartographers draw their dividing line between the Eastern and Western?

In 1884 by international agreement, mapmakers decided to draw the *prime meridian,* or 0 degree longitude, through the Royal Observatory of Greenwich, a borough of London, England. The reason was that at that time, the observatory had the most advanced technology for viewing and recording astronomical, meteorological, magnetic, and seismological phenomena. In other words, The English scientists knew the most about what *on Earth* (and above the earth) was happening.

So the longitudinal line dividing the earth into the Eastern and Western hemispheres was established to slice from north to south in a broad circular route—through the two poles, through Greenwich, and through a wide swath of the eastern Pacific Ocean, 180 degrees or half a circle away from Greenwich. (The line that runs exactly opposite the 0 degree prime meridian, or 180 degrees away on the other side of the earth, is called the *international date line.* The reason for this will become clear in the discussion below of time zones.)

For purposes of measuring time and longitudinal and latitudinal distances, the Western Hemisphere begins at the prime meridian, which runs through Greenwich, and ends 180 degrees west, in the Pacific Ocean on the other side of the world. For these purposes, the Western Hemisphere contains North and South America and the western parts of Europe and Africa.

The Eastern Hemisphere, according to this measurement, begins at the prime meridian and ends 180 degrees to the east, in the Pacific Ocean, where it meets the edge of the Western Hemisphere. This version of the Eastern Hemisphere contains Asia, Australia, the Middle East, and *most* of Africa and Europe.

But there's also another, more common definition of the Western and Eastern hemispheres that can be traced to traditions established in the European exploration of the Americas by Christopher Columbus and other adventurers in the fifteenth century. Beginning in those early days, Europeans referred to North and South America as being in the Western Hemisphere, and *all* of Europe and

Africa as being in the Eastern Hemisphere. Of course this custom began long before the 1884 conference that established Greenwich, England, as the site of the prime meridian. How can one make this earlier and still widely accepted definition work today from a car-tographer's point of view? The longitudinal line that separates the Eastern from the Western must be moved from Greenwich to the West. One common solution has been to regard the meridian at 20 degrees west as the dividing line on one side of the world, and the meridian at 160 degrees east as the dividing line on the other side. This way, maps that show the two hemispheres can place Europe and Africa entirely in the Eastern Hemisphere.

THE GEOGRAPHICAL TIME MACHINE

Geography is intimately connected with our understanding and measurement of time. You might say that an exquisite kind of geographical time machine has been created by the interplay be-tween the prime meridian and lines of longitude on the one hand, and the way you set your clocks on the other.

Here's the way this "time machine" works:

First, the foundation for the machine is rooted in some basic facts about the rotation of the earth. Earth makes one complete rotation around its axis between sunset one day and sunset the next day. We call one of these rotations a day.

Days are divided into twenty-four hours, and those twenty-four hourly segments correspond to twenty-four divisions indicated by certain lines of longitude on our maps. More specifically, one hour of time usually corresponds to 15 degrees of longitude. In other words, in most places on our planet it takes one hour for the rays of the sun shining at a certain angle to move a distance of 15 degrees on the earth.

By international agreement, all time begins at the prime merid-ian, or that line of 0 degree longitude that runs through Greenwich, England. In a narrow 15-degree band of longitude, which has Greenwich and the prime meridian at its center, all clocks are set at the same hour, minute, and second. The time in this zone is known as Greenwich Mean Time (GMT) or Universal Time (UT), and is calculated through astronomical observations.

As you move west, the time becomes one hour *earlier* for each 15 degrees of longitude. As you move east, the time becomes one hour *later* for each 15 degrees of longitude.

When the sun is directly overhead in any of these twenty-four time zones, the time is noon. In the hours before noon back until midnight, the time is designated as A.M., which stands for ante meridian, or before the meridian. In the hours after noon and extending forward to midnight, the time is stated as P.M., which means post meridian, or after the meridian.

Here are a few examples of how the different time zones relate to one another. Suppose the Greenwich Mean Time or Universal Time is noon. In New York City, it would be 7:00 A.M. Eastern Standard Time, because New York is five time zones (five 15-degree bands of longitude) to the west of Greenwich. In Hong Kong, the time would be 8:00 P.M., because Hong Kong is in a much earlier (more eastern) time zone band than Greenwich.

An important exception to this scenario: Over the years, inconsistencies have arisen in different parts of the world in the 15-degree standard distance delineating the zones. You'll find the time zone lines wandering in wild ways across the reaches of central China and Russia. Even in the continental United States, the four time zones from East Coast to West Coast don't stay precisely on the 15-degree longitudinal marker.

A Form of Time Travel?

Finally, what happens when you travel so far to the east that you pass the line of 180 degrees longitude? This 180-degree line coincides with the international date line, which marks the end of one day and the beginning of another. Suppose that the time is Wednesday noon in Greenwich. If you're out enjoying the moonlight just before midnight on a boat in the South Pacific, say off the shore of Western Samoa, to perform a very simple kind of time travel all you'd have to do to move back to Tuesday would be to take a few paddles to the east over the date line!

TRAVELER, KNOW YOUR SEASONS!

Usually, the wise, experienced traveler will peruse the international weather report before he leaves for a distant destination. But it's also smart to take into account the *season* of the destination area and the weather characteristics of that time of the year—especially if a relatively long stay is expected. A sense of the geography of the seasons in different parts of the world can make this sort of preparation much easier.

There are only a few basic facts you need to grasp to understand how geography influences the seasons of spring, summer, fall, and winter. First, you should remember that only two bands of temperate zones on the earth have pronounced variations among the seasons. The upper band, which encircles part of the Northern Hemisphere, is bounded by the Arctic Circle on the north (at 66.5 degrees north latitude), and the Tropic of Cancer on the south (at 23.5 degrees north latitude). The lower band, which lies below the equator, is marked by the Antarctic Circle on the south (at 66.5 degrees south latitude), and the Tropic of Capricorn on the north (at 23.5 degrees south latitude).

The main reason that we have seasons is that the earth remains in a permanent tilt of 23.5 degrees as it circles the sun during the course of a year. At the point in the earth's journey, when the North Pole is tilted most toward the sun, summer occurs in the Northern Hemisphere. At the same time, winter is going on in the Southern Hemisphere.

When the North Pole is tilted away from the sun, however, the reverse happens: There is summer in the Southern Hemisphere and winter in the Northern Hemisphere. Spring and autumn arrive when the earth is between its winter and summer positions in the path around the sun.

A word about equinoxes and solstices: Twice a year, the earth reaches a point in its orbit around the sun when the sun is precisely overhead at the equator. This happens on about March 21 and September 23 each year.

In the Northern Hemisphere, the March event is known as the spring equinox; the September event is the autumn equinox. At those times, the periods of light and dark are about equal during a twenty-four-hour day throughout the earth. (Note: The March and September events have the reverse designations in the Southern Hemisphere, with March 21 being the fall equinox and September 23 being the spring equinox.)

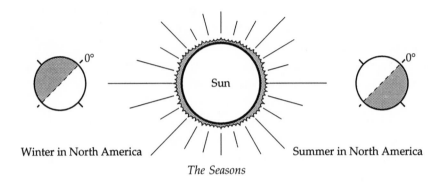

Winter in North America / Summer in North America

The Seasons

There are also two solstices. The summer solstice in the Northern Hemisphere comes on about June 22, and the winter solstice happens on about December 22. The summer solstice marks the longest period of daylight in a single day, while the winter solstice occurs on the day of shortest daylight. (Again, the designations of these events are reversed in the Southern Hemisphere.)

It's always helpful to keep these points in mind when you're traveling to a part of the world that is located at a different level of latitude from your starting place. A real problem may arise for a globe-trotting New Englander who assumes it's summer in Buenos Aires or Melbourne, Australia, just because it's summer in Massachusetts. I've even met otherwise savvy travelers who have gotten into trouble while traveling in the continental United States. One major mistake is to check only one recent weather report about your destination without taking into account broader seasonal issues.

For example, one woman had to fly in November from her home in San Antonio, Texas, to Bismarck, North Dakota. She took a cursory look at the most recent weather report in her newspaper and was pleasantly surprised to see that the temperature in North Dakota was relatively mild—only ten degrees or so cooler than in Texas. So she decided to pack only a light coat for her visit, which was scheduled to last several days. Unfortunately, the pleasant temperature she had noted was an aberration. The day after she arrived in Bismarck, there was a cold snap, which was typical for that time of the year. The temperature plummeted to below freezing, and she had to go out and buy extra clothing to stay warm.

What this woman didn't know was that the average temperature in Bismarck in November is typically less than 30 degrees Fahrenheit, while the average in San Antonio is about 60. It's true that it was autumn in both places—and both were in the temperate zone in the Northern Hemisphere. But her trip required covering nearly 20 degrees of latitude to the north, or almost half the distance from her home to the Arctic Circle! If she had possessed a better sense of how the seasons might vary at her destination at that time of the year because of the earth's tilt, she might have been more wary about relying on one isolated weather report at that rather changeable time in the earth's revolution around the sun.

Of course, an almanac can provide information about average climate conditions in different parts of the world. But you don't need to consult an almanac every time you are faced with a decision based on possible temperature variations. All that is necessary is a calculated, common-sense guess about the climatic conditions in light of the latitude of your destination and the season of the year when you're traveling.

With this overview of our globe in mind, let's turn now to a more detailed consideration of how the largest land masses—the continents—fit into the picture.

CHAPTER SEVEN

COMPREHENDING THE CONTINENTS

The seven continents are the cornerstones of world geography. The fifty-seven million square miles of dry land on our planet can be found on the continents—which are generally defined as the earth's major large land masses.

The shape, size, and position of the continents are constantly changing as a result of the movement of the earth's crust and underlying geological plates. But the rate of change is very slow: For example, Europe and North America are moving away from one another steadily, but at a snail's pace of only about an inch each year.

In order of physical size, with population estimates as well, here are the seven great land masses:

Asia, 17 million square miles, 3.1 billion people.

Africa, 11.7 million square miles, 625 million people.

North America, 9.4 million square miles, 420 million people.

South America, 6.9 million square miles, 290 million people.

Antarctica, 5.1 million square miles.

Europe, 4 million square miles, 710 million people.

Australia, 3 million square miles, 16.5 million people.

Any world map will show how the seven continents are positioned.

As you can see, the two continents of Europe and Asia are joined—a fact that explains why together they are sometimes referred to as the "Eurasian land mass." Europe is actually a vast peninsula of this land mass and is separated from Asia by the Ural Mountains and Ural River in the east.

Africa is connected to Asia only by the narrow strip of land that is cut through by the Suez Canal. North America and South America are also connected by relatively narrow ribbons of land in Central America. Australia and Antarctica look more like large islands. (You'll recall that Greenland is an island and Australia is not because Australia is significantly larger.)

HOW CAN YOU REMEMBER THE NAMES OF THE CONTINENTS?

Any geographically literate person *must* be able to name all seven of the continents, but for many people it's like naming the seven deadly sins. There's always a tendency to leave one or two out.

To help you hone your memory for the continents, try one of these two mnemonic devices:

- Arrange the first letters of the continents (but use North and South only; America isn't one of the *A*'s) this way: SANE-AAA. So long as you can remember all four that begin with *A*, this method should work fine.
- If you find you have trouble with the Four A's method, you might try this approach: Separate Europe from the others by thinking, "The name of this continent sounds like what you do if you're a cowboy: 'You rope.' "
- Then, imagine yourself "roping" the other six—all of which have a major word that begins with *A* and ends with *a*—according to where they are located with respect to the equator.
- So you rope two above the equator: Asia, and one of the Americas (North).
- You rope two on the equator: Africa, and the other America (South).

- You rope the last two below the equator: Australia, and Antarctica.

Now with the names and locations of the continents fixed firmly in your mind, let's move on to a closer consideration of each of these great land masses.

LET'S TAKE THE GRAND TOUR OF EUROPE

To acquire an in-depth knowledge of every country of Europe would require reading a library full of history books and tour guides. In this overview of the earth's geography, we obviously don't have the space for such a treatment.

But what we can do is highlight the major nations of Europe and identify a few "geographical keys" that every traveler and other well-educated person should know about each of these countries. A geographical key, as I'm using the term, is a fundamental fact or two that you *must* know about the physical features of each land. These keys are short and simple, but they are essential first steps as a basis for further knowledge, study, and understanding of the area.

The United Kingdom of Great Britain and Northern Ireland

Great Britain is the official name for the area that includes England, Scotland, and Wales. The entire nation, which also includes Northern Ireland, is known as the United Kingdom of Great Britain and Northern Ireland, or simply, the United Kingdom. In popular speech, the United Kingdom is sometimes referred to by its abbreviation, "the UK," and a popular name for all residents of the UK is the "Brits." The country is also known as the British Isles, or Britain. The capital is *London*, which lies near the southeast coast of England.

Caution: The name England is often (erroneously) used to refer to all of Great Britain. Remember: England occupies the majority of Great Britain on the southeast section of the island—but it's not the whole of Great Britain. If you use this designation in Wales or Scotland to refer to all of Great Britain or to the entire UK, you'll encounter a notable lack of enthusiasm.

Here are few more specific points on the non-English areas of the UK:

Northern Ireland is located in the northeast part of Ireland. It's an administrative unit of the UK and is closely bound to the rest of the nation by strong economic ties. Northern Ireland is also commonly called Ulster, and its capital is *Belfast*. Northern Ireland has a long history of national and religious conflict, with Protestants pitted against Roman Catholics in many pitched battles and terrorist incidents.

Wales occupies the western peninsula of Great Britain, with the capital at *Cardiff*. Welsh culture, from the Middle Ages on, has been known for its writers and singers.

Scotland is the northern portion of Great Britain and includes many surrounding islands. The capital is *Edinburgh*, and the largest city, *Glasgow*. Well-known symbols of Scotland are bagpipes, kilts, and the national emblem, the thistle.

The Republic of Ireland

This nation occupies most of the large island just to the west of Great Britain, with Northern Ireland occupying its upper tip. The country was proclaimed a separate, self-governing entity in 1949, with the capital and largest city being *Dublin*. Ireland is often called the Emerald Isle because of its brilliant green grass, which flourishes as a result of heavy rainfalls. Ireland's fine linen, handmade laces, and shamrock are world famous.

France

This country—located on the European continent just to the south of Great Britain, across the English Channel—has been a friend and ally of the United States since the latter declared independence.

France, with its capital at *Paris,* is especially known for wines, cheeses, and gourmet cooking.

Spain

With Portual, Spain forms the Iberian peninsula on the far western tip of the European continent. (Spain is just to the east of Portugal.) The capital is *Madrid.* A popular spectator sport and tourist attraction is bullfighting. The country is also known for olives, oranges, and Picasso.

Portugal

The capital of this country, which is the western-most nation in Europe, is *Lisbon.* As a neutral city during World War II, Lisbon became a center of international intrigue and political activity. Many spy movies were made in Lisbon. The city's outdoor cafés are favorite eating places for travelers, and the beach resorts south of Lisbon in the Algarve region are popular vacation spots. Portuguese bullfighting is also a popular attraction—but unlike the Spaniards, the Portuguese don't kill their bulls!

The Netherlands

Holland, which is the dominant province, is the popular name for the Netherlands. This country lies on the northern coast of Europe, just to the west of Germany and to the north of Belgium. The Hague is the seat of government but the constitutional capital is *Amsterdam,* renowned for diamond cutting, canals, tulips, and the great art museum, the Rijksmuseum. Almost half of the Netherlands is below sea level—hence the need for the famous dikes to hold back the waters of the North Sea.

Belgium

Situated just to the south of the Netherlands, Belgium is the headquarters of the twelve nations of the European Community and of the North Atlantic Treaty Organization (NATO). French is spoken in the south, Dutch in the Flemish north. The capital is *Brussels*. Known for its fine chocolates.

Luxembourg

Begun as a castle built in the tenth century on the site of an old Roman fort, this is one of Europe's oldest and smallest independent countries, bounded on the north and west by Belgium, on the east by Germany, and on the south by France. It's an international finance center, with its capital at the city of *Luxembourg*.

Germany

The nations of communist East and democratic West Germany were united in 1990 in one of the great watershed events in modern history. *Berlin* is now the capital. The nation had been divided in two since 1949. Germany, bordered by ten countries (France, Luxembourg, Belgium, the Netherlands, Denmark, Poland, Czechoslovakia, Austria, Liechtenstein, and Switzerland), continues to be a major economic power that will almost certainly become stronger and more influential in the years ahead. Known for Mercedes-Benz automobiles and beer.

Switzerland

Located just to the east of France and the north of Italy, Switzerland is known for the Alps, international banking, and the watchmaking industry. The government has usually assumed a neutral position in international politics and military actions. The capital of this beautiful, mountainous nation is *Bern*.

Italy

Italy, which includes many islands, the largest being Sicily and Sardinia, has its capital at *Rome*. This peninsular nation, positioned just to the south of Switzerland and France, is often said to be shaped in the form of a boot. It's known as the home of pasta, including spaghetti—and pizza—which, with certain national variations, have become staples of the diet of the United States and other countries.

Austria

This nation, with its capital, *Vienna*, is known to tourists for spectacular mountain scenery and winter sports, especially in the Tyrol region. Positioned to the south of Germany and Czechoslovakia and to the north of Yugoslavia and Hungary, Austria has gained popular recognition as the home of one of the great figures in the history of music, Wolfgang Amadeus Mozart. It is known also as the setting for the popular play and movie, *The Sound of Music.*

Poland

A country under the influence of the former Soviet bloc, Poland is bordered on the north by the Baltic Sea, on the west by Germany, on the south by Czechoslovakia, and on the east by Belarus and the Ukraine. Poland came to the forefront of the world's political stage in the 1980s as the independent Solidarity movement gained momentum. Solidarity, which gained control of the government in 1989, refers to a labor union movement that has demanded greater democratic control of industry. The capital of Poland is *Warsaw.* Known widely for the export of ham.

Czechoslovakia

This land, located to the south of Poland and Germany, is in the process of peaceful division, and the future depends on the results of negotiations between the Czech and Slovak National Councils. The capital city is *Prague.*

Hungary

This country lies to the north of Yugoslavia, to the east of Austria, and west of Romania. Known for the highly seasoned stew, called goulash. The capital is *Budapest,* which lies on both banks of the Danube River.

Yugoslavia

This predominantly agricultural country has *Belgrade* as its capital city. To the west is the Adriatic Sea, to the east Romania and Bulgaria, to the south lies Greece. For details on the recent breakup and conflicts in Yugoslavia, see Chapter Four.

Albania

This rugged, mountainous nation has a fertile coastland on the Adriatic Sea, just to the northwest of Greece. Albania has been a fiercely independent communist country, with the lowest standard of living on the continent of Europe. The capital city is *Tiranë.*

Greece

Greece, which rests on a peninsula to the south of Albania, Yugoslavia, and Bulgaria, is bounded on the west by the Ionian Sea, on the south by the Mediterranean Sea, on the east by the Aegean Sea, and on the northeast by Turkey. The nation includes many islands, the largest being Crete and Rhodes. Ancient Greek culture was the main source of Western civilization. Current major produce includes olives, olive oil, wine, and, in America, coffee shops!

Bulgaria

Part of the region controlled by the old Soviet Bloc, this country was industrialized by the former USSR, but it still remains mostly agricultural. A nation bordered by Yugoslavia, Romania, and Greece, with the Black Sea to the east. The capital is *Sofia.*

Romania

Another former Soviet Bloc country directly north of Bulgaria, Romania is now independent with the capital at *Bucharest.*

Russia

The western part of Russia, up to the Ural Mountains and including *Moscow* (the capital), is considered to be the eastern-most part of Europe. The area of Russia east of the Urals is in Asia. Known for its ballet dancers, caviar, and vodka. A common mistake: Many people assume that all of Russia is in Asia—but that's not correct. For further details on Russia, see Chapter Two.

Turkey

Only the western-most 3 percent of this nation is in Europe. The European portion is just to the west of the Bosporus Strait and the Sea of Marmara, and to the east of Greece and southeast of Bulgaria. The rest is in Asia. *Ankara,* the capital, lies in Asia, but the largest city, Istanbul, is situated *both* in Europe and Asia! (The site of ancient Troy is believed to be on the western coast of Asiatic Turkey.) Major export is carpets.

Latvia, Lithuania, and Estonia

These Baltic states, which were part of the old USSR, are described in detail in Chapter Two. The capitals of these three republics are *Riga, Vilnius,* and *Tallinn,* respectively.

Liechtenstein

This is one of the smallest and wealthiest countries in Europe, with sixty-one square miles and a population of thirty thousand. The nation is bordered on the west by Switzerland and on the east by Austria. The capital is *Vaduz*.

Norway

Known everywhere for its fjords, or narrow sea inlets surrounded by steep cliffs, and its reindeer, this Scandinavian nation has been nicknamed The Land of the Midnight Sun. The reason: About one third of the country lies above the Arctic Circle, so the sun stays above the horizon all day and night during the summer. (But it can't be seen at all in the winter!) Norway snakes along the western border of Sweden and wraps around the top of Sweden and Finland. The capital city is *Oslo*. Well-known exporter of sardines.

Sweden

This Scandinavian country has the world's most comprehensive social welfare programs—and also some of the best tennis players. The Nobel Prizes are awarded in the capital, *Stockholm*.

Iceland

Iceland is an island country in the North Atlantic just south of the Arctic Circle, southeast of Greenland, and west of Norway. It proclaimed its independence from Denmark in 1944. Capital and largest city is *Reykjavik*.

Denmark

This peninsular Scandinavian nation, which includes many islands, is tacked onto the north of Germany and juts out into the North and Baltic Seas. Denmark, with the capital at *Copenhagen*, has long and illustrious literary connections: Shakespeare's Hamlet was the Prince of Denmark, and the country was also the home of the storyteller Hans Christian Andersen.

Finland

Finland, also one of the Scandinavian countries, lies just to the east of Sweden, across the Gulf of Bothnia, and to the west of Russia. The capital is *Helsinki,* site of many international conferences.

Ukraine, Moldova, Belarus (Byelorussia)

These new nations, which became independent after the breakup of the USSR in 1991, are also located in Europe, just to the west of Russia. For more details, see Chapter Two. Their capital cities are *Kiev, Kishinev,* and *Minsk,* respectively.

ALL YOU NEED TO KNOW ABOUT ASIA

Asia is the world's largest continent, with 30 percent of the surface land area and three fifths of the earth's population. It's estimated that the population of this huge continent may reach 3.4 billion people by the year 2000.

But as big and important as Asia is, mistakes are always being made about what lands and nations lie within its orbit. For example, many people don't realize that the vast, sandy deserts of Saudi Arabia are part of Asia, not Africa. Also, as I've already mentioned, western Russia and its capital Moscow are *not* in Asia, while most of Turkey is. Even the eastern tip of Egypt, the Sinai, is in Asia, not Africa.

Asia reaches from Israel, to Indonesia and New Guinea, to Siberia. Its boundary with Europe lies approximately along the Ural Mountains, to the east of Moscow in Russia, and runs down along the Ural River and the Greater Caucasus.

The lands of Asia include the countries of the Middle East, an area that has been described in some detail in Chapter Two. Asia also encompasses the eastern part of Russia, all of China, and Japan. Other nations that are part of the Asian continent may be part of the mainland or they may be islands or archipelagoes (clusters of islands): Afghanistan, Pakistan, India, Sri Lanka, Mongolia, Nepal,

Bhutan, Bangladesh, Malaysia, Singapore (a tiny island and nation, as well as a city), Brunei, Indonesia, Burma, Laos, Cambodia, Thailand, Kampuchea, Vietnam, North Korea, South Korea, Taiwan, and the Philippines.

As you can see, Asia, with all its island nations, is more an amorphous geographical hodgepodge than the homogeneous, continuous land mass we normally associate with the word *continent.* It's the most physically diverse of all the seven continents, with the greatest range of land height, climate changes, vegetation, and animal life.

To understand just how varied this great continent is, consider the basic data included for each of the following Asian nations. I've chosen to highlight these countries because they are the ones that tend to appear most often in news or other written accounts—and they are also among the most frequent travel destinations for business people and tourists.

Japan

The capital of Japan (or *Nippon,* in Japanese) is Tokyo, one of the world's largest, most modern cities. The population of Japan, which is known as The Land of the Rising Sun, is more than 124 million. Japan's flag, with a red circle on a white background, symbolizes the sun image. Japan is *not* the name of one island, but rather of an archipelago, or group of islands. The islands that make up Japan include four large ones: Honshu (the largest, which contains Tokyo), Hokkaido (the northern-most), Kyushu (the southern-most), and Shioku. These four and about three thousand (!) smaller ones are separated by the Sea of Japan from China, Korea, and Russia.

One of the best known geographical landmarks in Japan is Mount Fuji, or Fujiyama, a volcano that last erupted in 1707. Mountains, most of them volcanoes, cover about two thirds of Japan's surface. Less than 20 percent of the land is cultivated—a fact that drives the Japanese to develop their industries and manufacture products that may be exchanged for food and other natural resources from other nations. The temperatures vary from chilly to warm; rainfall is abundant; and typhoons and earthquakes are frequent.

Japan, as mentioned in an earlier chapter, is on the unstable, earthquake- and volcano-prone region called the Ring of Fire, which encircles the Pacific. The San Andreas Fault in California is part of the Ring and is very much alive geologically.

Tokyo, the largest urban and industrial center in the country, is served by a dense network of electric railways, subways, bus lines, and highways. The high-speed bullet trains zip between Tokyo and other major cities, such as Osaka, the second-largest city. Osaka was the capital in the sixteenth century.

People's Republic of China

Also known as Communist or mainland China, this nation has the largest land mass in the world after Russia and Canada. *Beijing* is the capital. Since 723 B.C. several cities, bearing various names, have existed on the same site. Shanghai is the largest city.

China was ruled by emperors until the early part of the twentieth century. Then, in 1911, the last emperor was overthrown and a republic was established. The communist movement gained strength beginning in the 1920s, civil war broke out, and the present communist regime took control of the government in 1949. Current social and political conditions are dynamic, with pockets of free enterprise springing up and Western influences gradually increasing. The ultimate response of the government to these changes remain uncertain, however.

Hong Kong

This British colony and world trade and financial center, which consists of Hong Kong Island and Kowloon Peninsula, is located on the south coast of China. The territory was leased to Britain in 1898 for ninety-nine years. With the signing of the Chinese-British joint declaration in 1984, both countries agreed that the territory would be returned to China on July 1, 1997. Hong Kong has been described variously as a geographical "jewel" because of its beautiful physical features; a paradise for shoppers, especially those interested in inexpensive but high-quality handmade clothing; and a center of Asian nightlife.

India

The Republic of India, which is Asia's next-largest country, after Russia and China, fills the major part of the Indian subcontinent, on the southern part of the Asian mainland. Other nations that share this subcontinent include Pakistan, Nepal, Bhutan, Bangladesh, Sikkim, and Sri Lanka. India's northern boundary with Asia is the high Himalayan mountains, which are the home of the world's highest peak, Mount Everest (29,028 feet above sea level on the border between Nepal and China).

Because the country is so vast, yet central in joining the East with the West, it has many cities which are both unique and important. Here are some basic facts about India's major cities:

New Delhi is the capital of the country. It was built between 1912 and 1929 to replace Calcutta as the capital of British India. The southern section of the city harbors the prayer ground, where the great political leader and pacifist Mahatma Gandhi was assassinated in 1948.

Calcutta, located on the upper east coast of the country, is the largest city in India and one of the largest in the world. It suffers from severe overcrowding and is the place where Mother Teresa of Calcutta has based her Christian ministry to the poor. Calcutta was also the site of the "black hole of Calcutta." This was the cell in a British jail that was used by Indian troops, under the leadership of the Nawab of Bengal, to imprison 146 British defenders. Most of them died of suffocation. The city was developed on an English factory site established in 1690 and became the chief port of East India.

Bombay, the second-largest city in India, is the main port of West India and the only deep-water harbor on the western coast. Archeological remains on the site go back to 320–184 B.C., when the city was part of an ancient Buddhist empire.

EXPLORING AFRICA

If Asia projects an exotic image to Westerners, Africa conjures up a sense of mystery, largely because some of the greatest exploring expeditions of the nineteenth and early twentieth centuries were launched into this wild and exciting land. The Stanley–Livingstone encounter is just one example that comes to mind.

Africa, the second-largest continent, occupies one fifth of the land area of the earth. One quarter of this land is desert—the Sahara. From north to south, the continent is cut almost equally in half by the equator. As a result, most of Africa lies within the tropical region between the Tropic of Cancer (at 23.5 degrees north latitude) and the Tropic of Capricorn (at 23.5 degrees south latitude).

Off the coasts, a number of islands are also linked to the continent. The largest of these is Madagascar, which lies off the southeast coast. This huge island is the home of those fascinating creatures, the lemurs.

The greater part of this continent has always been inhabited by black peoples, but there have also been major immigrations from both Europe and Asia. The people of Africa probably speak more languages than those of any other continent—an astounding eight hundred!

The highest point on the continent is Mount Kilimanjaro in Tanzania (19,340 feet above sea level), which is always snow-capped, despite its nearness to the equator. Africa's climate is dominated by its position astride the equator. Temperatures are high for most of the year, though they are modified by elevation in the mountains and by the influence of ocean currents near the coasts. Africa is well known for its exotic animals, including the elephant, rhinoceros, giraffe, zebra, lion, buffalo, leopard, gorilla, and cheetah. Big-game animals are found roaming in the savanna regions, and some of the world's finest national parks were originally established as game reserves in Kenya, Uganda, Tanzania, Zambia, and South Africa.

In many ways, Africa may be our most ancient continent. Many anthropologists believe that the first humans lived here. Africa was also a center for some of the earliest civilizations, including the first great kingdom of Egypt, which arose along the Nile beginning approximately in 3000 B.C. With the Sumerian kingdom of Mesopotamia, the early Egyptian culture can lay joint claim to being a "cradle of civilization."

Today, Africa is divided into fifty-five different countries. Here are some highlights:

Algeria

With the capital at *Algiers*, this nation is 85 percent desert. Almost the entire population lives in the northern part—known as the Tell; the remainder of the country is uninhabited, except for oases that punctuate the arid region.

The country is situated on the south coast of the Mediterranean Sea, between Libya on the east and Morocco on the west. Algeria was colonized by France in the nineteenth century and gained its independence in 1960.

Libya

Libya, which lies on the northern edge of the continent, between Algeria and Egypt, is another desert land, with 99 percent of its surface covered by a Saharan plateau. The capital is *Tripoli*, the site of battles by the United States Marines against the Barbary pirates. These encounters inspired the "shores of Tripoli" reference in "The Marines' Hymn." Numerous oases, watered by wells and springs that are fed by an underground water table, relieve the dryness. The climate is generally hot, but more comfortable temperatures characterize the shores along the Mediterranean Sea.

The Sudan

The Sudan, positioned just to the south of Egypt, is the largest African country; its capital is *Khartoum*. The north is rock desert, with an average annual rainfall of zero inches. As you go south, however, the rain increases to almost fifty-seven inches annually in the extreme south. The dominant geographical feature is the Nile River, which is formed by the confluence of the White Nile and the Blue Nile rivers. Unlike most rivers, the Nile flows north. The Sudan gained its independence in 1956. Prior to that, it was under the joint rule of Britain and Egypt.

Ethiopia

Sometimes called Abyssinia, this nation, located to the southeast of the Sudan and to the west of the Red Sea, has been in continuous existence for about two thousand years. The capital is *Addis Ababa.* Ethiopia became an outpost of Christianity in the fourth century, but a large minority of Ethiopians are now Muslim. The last emperor in Ethiopia was the powerful Haile Selassie. A military takeover in 1974 deposed him and turned the country into a socialist republic.

Kenya

Positioned just below Ethiopia on the eastern coast of the continent, Kenya has part of its border on the Indian Ocean. The capital is *Nairobi.*

The country is renowned for its abundant wildlife and has set aside vast preserves as protected territory for endangered species. Kenya's Great Rift Valley is the site of some major archeological discoveries, including the remains of the earliest known humans. The beauty and variety of the landscape, as well as a pleasant and sunny climate, have made it attractive to writers and motion picture companies. Works set in Kenya include *The Green Hills of Africa* by Ernest Hemingway, *Out of Africa* by Isak Dinesen, and *Born Free* by Joy Adamson. The publicity from these books and movies, and the colorful life-style of the internationally known Masai tribe, have drawn more tourists to Kenya than to any other African country.

Tanzania

With the capital at *Dar es Salaam*, this country lies just to the south of Kenya on the Indian Ocean. The nation was formed in 1964 by the union of Tanganyika and the island state of Zanzibar.

At Olduvai Gorge in northeastern Tanzania, Louis B. Leakey, a British anthropologist, found the 1.75-million-year-old remains of what he claimed was a direct ancestor of the present human species. Tanzania teems with animal life, including the five major big-game animals: elephant, rhinoceros, buffalo, lion, and leopard. The national parks are some of the finest in Africa.

Zaire

With the capital now at *Kinshasa,* this country was formerly called the Belgian Congo. It occupies the center of the Congo River basin, in the heart of Central Africa. (Remember Joseph Conrad's "Heart of Darkness"?) The Congo is one of the world's major rivers and one of the longest. Zaire also has the most extensive rain forests in Africa, and is the home of gorillas, elephants, baboons, okapi, giraffes, lions, and cheetahs. A country rich in economic resources, Zaire has vast deposits of industrial diamonds, cobalt, and copper.

Nigeria

Nigeria, which has its capital at *Lagos,* has the largest population of all African countries. Located just to the south of Niger on the South Atlantic Ocean in the "crook" or curved-in section of western Africa, Nigeria became independent in 1960. But it elected to stay a member of the British Commonwealth. The official language is English. Since there are more than two hundred different languages spoken by the many ethnic groups living in the country—and any consensus for changing things seems unlikely—the chances are that English will *remain* the official tongue.

Along the entire coastline lies a belt of mangrove swamp forest from ten to sixty miles wide. Beyond that is tropical rain forest, and then a plateau region takes over.

South Africa

This nation, which lies on the southern tip of the continent, has *three* capital cities: *Cape Town* (legislative capital); *Pretoria* (administrative capital), and *Bloemfontein* (judicial capital).

As the southern-most state in Africa, South Africa resides in a temperate climatic zone, a fact that contributed to European settlement on a scale unknown elsewhere in Africa. Prior to 1991, the government had maintained a policy of apartheid ("apartness"), which insisted on segregation and political and economic discrimination against non-European groups. This state of affairs evoked vehement opposition from most countries in the world, and in December of 1991 the policy of apartheid was abandoned. Change is finally in the air. Johannesburg is the largest city, which has grown at a remarkable speed since 1886, when gold was discovered in the area. This resource has made the city an economic metropolis and the center of the country's gold-mining industry.

Madagascar

This separate nation occupies the fourth-largest island of the world, off the southeast coast of Africa. The capital is *Antananarivo*. Originally covered with evergreen and deciduous forests, the country lost these resources through slash-and-burn farming methods. Now, there are only a few patches of forest, which are the home of about forty species of lemurs. These unique animals are found nowhere else in the world, primarily because of Madagascar's isolation. The island's population consists of eighteen to twenty tribal groups.

THE NEW WORLD OF NORTH AMERICA

Americans—including all those who live in North America, from Canada on the north to Panama on the south—live in one of the most developed regions of the world. We comprise less than 10 percent of the world's population, but we produce more than one third of all the manufactured goods on earth. The region's great natural wealth has accounted for much of its prosperity. Our continent offers extensive and seemingly inexhaustible deposits of metals and fuels, vast forests, ample water resources, and a wide range of climates and soils.

Most Americans learned in school that the name *America* derived from the Italian merchant and navigator, Amerigo Vespucci, one of the earliest explorers of North America. The part of the land that widens out north of the Isthmus of Panama came to be designated as North America. The name South America is applied to the lands that extend below Panama. But sometimes, the whole of Mexico and Central and South America are referred to as Latin America. The United States and Canada may be referred to as Anglo-America. This distinction reflects the pronounced cultural divisions, but Mexico and Central America are bound to the rest of North America by strong ties of history and physical geography.

The general terms *New World* and *Western Hemisphere* are usually employed to refer to North and South America, the Caribbean islands, and Danish-speaking Greenland. (As mentioned earlier, when *Western Hemisphere* is used this way, the earth must be divided in half at approximately the 20th and 160th lines of longitude.)

Take a look at your world to get a broad view of the lands encompassed by North America. You should refer to this map as you consider some of the major nations of North America.

Canada

Canada consists of all the North American continent north of the United States *except* for Alaska and Greenland; the latter is currently ruled by Denmark. (There are ongoing efforts among Greenlanders to make their island, which is the largest in the world, independent.) There are also two very small islands just off the south coast of the Canadian province of Newfoundland, Saint Pierre and Miquelon, which are classified as departments of France.

Note: Iceland, which lies to the east of Greenland, is considered to be a part of Europe.

The total area of Canada is 3.85 million square miles, which makes it the second-largest country in the world, after Russia. The northern islands comprise a large group extending up from the mainland to within 550 miles of the North Pole. Not too much research is needed to arrive at the conclusion that it's pretty cold up there! In fact, almost 90 percent of Canada has subarctic to arctic climatic conditions.

You may remember the old Laurence Olivier film called *The 49th Parallel,* about some Nazis who were on the run after their submarine sank in Hudson Bay. They had to be caught before they escaped across the border into the United States, which had not yet entered World War II and was therefore neutral territory. What is the significance of the 49th parallel? This line of latitude marks the western boundary between the United States and Canada.

Although precise census figures are not yet available, the population estimate for Canada as of mid-1992 is 27,400,000. Compare that figure with the mid-1992 figure of 254,105,000 who live in the United States! Canada is divided into provinces and territories. These include New Brunswick, Newfoundland, Prince Edward Island, Nova Scotia, Ontario, Quebec, Alberta, Manitoba, Saskatchewan, British Columbia, Northwest Territories, and Yukon Territory.

Now here are some things you should know about important Canadian cities:

Montreal: This city in Quebec province is the second most populous city in Canada. It has a major seaport on the St. Lawrence River, which connects with the Great Lakes area and is open to 80 percent of the world's maritime fleet. Both English and French are spoken throughout Montreal.

Ottawa: The capital of Canada, this large metropolitan area lies on the border between Ontario and Quebec. Originally named Bytown, it was rechristened *Ottawa* after the Indian tribe of the same name when it became capital of United Canada by order of Queen Victoria in 1855.

Quebec: This capital of Quebec province lies at the confluence of the St. Lawrence and St. Charles Rivers. The majority of residents are French-speaking, and the city maintains a dual school system, one for Catholics and one for Protestants. Instruction is in French for the Catholics and English for the Protestants.

Vancouver: This prosperous port, with access to the Pacific, lies just north of Washington state in British Columbia. The economy of Vancouver was stimulated greatly through the opening of the Panama Canal in 1915, an event that made it feasible to export grain and lumber from Vancouver to the American East Coast and Europe. Without the Canal, the Canadian ships would have had to sail around the bottom of South America, close to the South Pole, and then up the other side! Vancouver's atmosphere is somewhat British, with Oriental overtones. It includes a Chinatown that many consider second only to San Francisco's.

Toronto: The capital of the province of Ontario, this city is the most important and populous in Canada's most prosperous province. It has access to Atlantic shipping via the St. Lawrence Seaway and to major United States industrial centers, via the Great Lakes.

The United States

As the foremost country in the Western Hemisphere in population and economic development, the United States is a federal republic composed of fifty states. The total area is 3.6 million square miles, a size that makes it the fourth-largest country in the world, after Russia, Canada, and China.

The United States is bordered on the north by Canada; on the east by the Atlantic Ocean; on the south by the Gulf of Mexico and Mexico; and on the west by the Pacific Ocean. The highest peak in the United States, as well as in North America, is Mount McKinley in Alaska (20,320 feet above sea level); the lowest point is Death Valley, California, which is 282 feet below sea level. The physical environment of the United States ranges from the Arctic to the subtropical; from moist rain forest to arid desert; from bald mountain peaks to flat prairies. In other words, "from the mountains to the prairies, to the oceans white with foam . . ." as the song "God Bless America" goes.

Natural resources, though increasingly depleted and seriously polluted in some areas, continue to sustain an economic life that is more diversified than any other on Earth. Furthermore, the natural riches of the nation provide the majority of its people with an exceptionally high standard of living. Although comments on some of the major cities in the United States will be included in the next chapter, here is some basic information on several of the largest and most prominent cities in the country.

Washington, D.C. (District of Columbia): The capital of the country is situated at the navigational head of the Potomac River, between Maryland to the northeast and Virginia to the southwest. The District of Columbia was chosen by Congress in 1790 as the site for the permanent seat of government for the new nation. Washington thus became one of the few cities of the world planned expressly as a national capital. The site was proposed by George Washington and named after him.

New York: This port city is the largest metropolis in the United States and is widely regarded as the financial and cultural capital of the world. The center of the communications industry and major stock markets, such as the New York Stock Exchange, New York has been a strong magnet for immigration from all parts of the world since it was established by the Dutch as New Amsterdam in the seventeenth century. There continues to be tremendous ethnic diversity in the city.

Los Angeles: The hub of much of the film and television entertainment industry, this Southern California city is the second most populous in the United States, and sprawls over nearly 500 square miles. Hollywood is a district of Los Angeles.

Chicago: As the third largest city, Chicago is the dominant metropolitan area in the upper central region of the United States. Located in the state of Illinois on the coast of Lake Michigan, Chicago has access to trade routes that pass through the Great Lakes and up into Canada.

Houston and *Dallas, Texas,* and *Atlanta, Georgia,* are large urban centers that exert the most powerful cultural and economic influence in the south central and southern part of the nation.

Mexico

This "south of the border" nation is geographically positioned in the earth's great desert region. No major river crosses the country, and water is scarce throughout most of the territory. In many ways, Mexico is a bridge between the broad North American continent and the tropical features of Central America and the Caribbean region. Mexico includes the narrow peninsula of Baja California, which parallels the west coast of the mainland. The capital, *Mexico City*, is both the political and economic center of the country and a focus of Latin American culture since the sixteenth century.

Central America

This southernmost region of North America lies between Mexico and South America. Nations in this area include Guatemala, El Salvador, Belize, Honduras, Nicaragua, Costa Rica, and Panama. The region is flanked by the Caribbean Sea on the east and the Pacific Ocean on the west. Humid swamp and lowlands extend along both the west and east coasts. The nations in this part of the world are known as the "banana republics" because of the prevalence of that crop.

Guatemala: The capital, *Guatemala City*, is on a tropical plain averaging thirty miles in width and paralleling the Pacific Ocean. It has some thirty volcanoes, of which six have erupted in recent years. A catastrophic earthquake in 1976 left nearly twenty-three thousand dead, seventy thousand injured, and one million homeless. Guatemala's main environmental problem is deforestation. More than 50 percent of the forests have been destroyed, and extensive soil erosion has resulted. The country has a particularly low average income and literacy rate and a rather unstable government.

El Salvador: With the capital at *San Salvador*, this is the smallest Central American country. It's a land of mountains and once-fertile upland plains before deforestation caused soil erosion. There is no enforced national law to protect the environment. To add to these troubles, the country is constantly torn by civil unrest and terrorism. The main export is coffee.

Belize: Formerly known as British Honduras, this small nation, with the capital at *Belmopan,* is mostly level land with a subtropical and humid climate. Because of its sparse population, Belize encourages immigration. The official language is English, but nearly everyone speaks a Creole patois.

Honduras: The capital of this mostly mountainous nation is *Tegucigalpa.* Only the southern coastal plain and land around two river valleys is relatively flat. One of the three largest exporters of bananas in the world, Honduras is nevertheless the poorest nation on the American mainland. Tourism is growing, however, with the main attraction being the restoration of ruins at Copán, the second-largest city of the ancient Mayan empire.

Nicaragua: This is the largest Central American country, with the capital at *Managua.* The Caribbean coast, known as the Mosquito Coast, consists of low, flat, wet, tropical jungle extending inland for fifty to one hundred miles. The country also lies in an earthquake zone, and the last major quake, in 1972, claimed ten thousand lives.

Costa Rica: With the capital at *San José,* this country has three main topographical regions—highland, plateau, and lowland. The lowlands have an average annual rainfall of one hundred inches, with almost continuous rain along the Caribbean side. The main exports are bananas and coffee. Considered perhaps the most politically stable country in Latin America, Costa Rica has a literacy rate of over 90 percent. It's traditionally been democratic politically.

Panama: This southernmost country of North America is situated on the Isthmus of Panama (an isthmus is a narrow strip of land connecting two larger land masses). The capital is *Panama City.* A major man-made feature of the country is the Panama Canal, which connects the Caribbean Sea (and the Atlantic Ocean) with the Pacific Ocean. The Canal Zone, over which the United States formerly exercised sovereignty, was incorporated into Panama on October 1, 1979. The United States retained responsibility for operation of the

Canal and the use of land in the zone for maintenance of the Canal until the year 2000.

Panama is a country of heavily forested hills and mountain ranges. The Panama Canal utilizes a gap in these ranges. Both coasts of the isthmus have deep bays, and the Gulf of Panama, on the west side of the country, has good deep water anchorages. Panama has a tropical climate, but temperature varies according to altitude, with hardly any seasonal change in temperature. There are warm days and cool nights throughout most of the year.

The West Indies

Sometimes called the Antilles, these islands are scattered throughout the Caribbean Sea and form an archipelago more than fifteen hundred miles long between North and South America. The West Indies, apart from the Bahamas and Bermuda, are commonly divided into two groups: the Greater Antilles, and the Lesser Antilles. The Greater Antilles comprise the large islands of Cuba, Jamaica, Hispaniola (Haiti and the Dominican Republic share this island), and Puerto Rico. The Lesser Antilles include the Virgin Islands and the small islands that form the Windward and Leeward groups—Barbados, Trinidad, and Tobago. There is also a group of islands called the Netherlands Antilles in the Caribbean Sea off the northern coast of Venezuela. The principal islands are Aruba, Bonaire, and Curaçao, all administered by the Netherlands.

The Bahamas and Bermuda, though not part of the West Indian region, have historic and cultural ties to the other islands.

The whole West Indian region is popular with travelers and tourists because of the warm seas, green mountains, and fertile valleys, which give the islands a quality often described as the nearest thing to paradise. But hurricanes are a major menace to the West Indian climate. Violent storms may blow in from the Atlantic and strike one or more of the islands at once. Then they can turn north across the larger islands toward North America.

Cuba is the largest island country in the West Indies, and accounts for more than half the West Indian land area. At this writing, Cuba remains an isolated, hard-line communist regime under the aging dictator Fidel Castro. But many feel that dramatic political and social changes are inevitable in the near future—a prospect that will almost certainly open this nation up to an exciting influx of tourism and trade.

THE ROAD TO RIO: THE CONTINENT OF SOUTH AMERICA

South America, the southern continent of the Western Hemisphere, comprises about one eighth of the earth's land area. This continent is shaped somewhat like a triangle, broad to the north and tapering to a point in the south. It's separated from Antarctica by the Drake Passage, and in the north, it's joined to North America by the Isthmus of Panama.

An outstanding physical feature of South America is a striking mountainous division between the eastern and western sections of the continent. The Andes mountains border the Pacific and run the entire length of the continent, or about five thousand miles. They create a gigantic backbone that divides the continent into two parts, which differ greatly in size and character. Because of the position of the mountain ranges on the west, rain that falls only a hundred miles east of the Pacific will flow down the slopes to the Atlantic twenty-five hundred miles away! The waterways of South America are dominated by the Amazon River basin, which drains a third of the continent's land area into the Atlantic Ocean. Two other important river systems are the Orinoco and the Río de la Plata. Most of the lakes of South America are mountain lakes in the Andes. Lake Titicaca, lying between Peru and Bolivia, is the highest freshwater lake in the world, at an elevation of 12,500 feet. More than half the land area of South America is covered by forest—principally the enormous, but rapidly diminishing Amazonian rain forests. The forests have been ravaged as the ground has been cleared for sugarcane plantations. About twenty-five hundred different species of trees grow in the rain forests of the Amazon alone. Almost a fourth of all known species of animals live in the rain forests, plateaus, rivers, and swamps of South America. They include such rare creatures as llama, jaguar, alpaca, sloth, giant anteater, manatee, and piranha.

Exploration of South America began in the sixteenth century, with the Portuguese claiming what is now Brazil, and the Spanish claiming most of the remaining land. The influence of these colonizing nations is still present in a number of respects, including the language: Portuguese is spoken in Brazil, and Spanish is spoken in the lands settled by the Spanish. All of the Latin American nations, however, have achieved their independence from Spain and Portugal. Though South America is becoming more and more urbanized, impoverished subsistence farming is prevalent.

The continent and its adjacent islands are divided into twelve independent countries: Argentina, Bolivia, Brazil, Chile, Colombia, Ecuador, Guyana, Paraguay, Peru, Suriname, Uruguay, and Venezuela. There are also two dependencies: The Falkland Islands, off the southeastern tip of the continent, are owned by Britain but claimed by Argentina. French Guiana is an overseas "department" of France, and its inhabitants are French citizens.

Now, here are some more details about these South American nations and dependencies.

Argentina

This Spanish-speaking country, with its capital at *Buenos Aires,* occupies most of the southern portion of South America. It's the eighth-largest country in the world and stretches from the subtropical north to the near-Antarctic south. In the nineteenth century, this former colony of Spain was the land of the gauchos, or the "lone horsemen of the pampas," and of ranchers who lived like kings on estates the size of small countries. The situation changed in the twentieth century when millions of poor Europeans came to Argentina to seek a decent living on the huge farmlands of the interior. They brought skills that helped transform Argentina into a modern country, with advanced agriculture and industry.

Bolivia

This country has two capitals: *La Paz* (the administrative capital) and *Sucre* (the judicial capital). Bolivia is a landlocked country just to the north of Argentina, in central South America. Once part of the ancient Inca empire, Bolivia is now an underdeveloped nation whose economic life is based principally on the production of tin. The government has tried to fight the widespread production and traffic of drugs, but with little success.

Brazil

The capital of Brazil is *Brasilia,* a planned city founded in 1957 which replaced Rio de Janeiro as the capital in 1960. Brazil occupies nearly half the continent of South America and is exceeded in area only by Russia, China, Canada, and the United States. It's made up of the former colonies of Portugal, but unlike the Spanish colonies, which became separate countries, the Portuguese colonies united into one huge country. Brazil is the world's leading coffee exporter.

Chile

With the capital at *Santiago,* this long, "skinny" country runs along the southern Pacific coast. Chile has an average width of slightly more than a hundred miles and a length of twenty-seven hundred miles! The nation is mainly mountainous. Because of its extreme length, Chile has a wide variety of climates, from desert in the tropical north to the cold, Antarctic-like tip. Chile's economy is based on agriculture and mining, but unlike many other Latin American countries, it also has developed a manufacturing capability. Thus, Chile is one of the more urbanized South American societies.

Colombia

Located at the northwest corner of South America, this nation, which has its capital at *Bogotá,* is a remarkable study in contrasts, both geographically and socially. The snow-capped mountains of the interior tower above the equatorial forests and savannas, where native tribes still follow the ways and traditions of their Stone Age ancestors. At the cooler intermediate elevations, modern cities have been built amid traditional rural landscapes, where small crops of coffee and corn are grown by native farmers. Political instability is closely tied to the unequal distribution of wealth. Illicit trade in drugs has also become a major disruptive factor in the economy.

Ecuador

Ecuador, whose capital is *Quito,* derives its name from the fact that it's crossed by the Equator. Although rich in natural resources, this country has not sustained a high rate of economic growth. The economy was basically agricultural until extensive exploitation of petroleum deposits began in 1972. This change stimulated the industrial development of the nation; however, the gap between rich and poor remains vast.

Guyana

The capital of this country, which lies on the northern coast of the continent and is a member of the British Commonwealth, is *Georgetown.* The economy is dominated by the sugar industry, and there is also a growing bauxite industry. The population is small, and for the most part, the people are descendants of slaves and indentured laborers, who were imported to work the sugar plantations.

Paraguay

With the capital at *Asunción,* this land-locked country is just to the south of Bolivia and to the west of Brazil. The Paraguay River and other rivers play a vital role in the nation's economic life by providing the country with access to the distant Atlantic Ocean and with sites for hydroelectric plants. The name of the country is derived from an Indian word that means "place with a great river."

Peru

Peru is located on the South Pacific coast just to the north of Chile. The capital is *Lima*. The nation's name is derived from a Quechua Indian word meaning "land of abundance," a reference to the highly organized and fruitful Inca civilization that ruled the region from the twelfth to the sixteenth centuries. The remains of this empire centered at Cuzco included the fabled stone fortress of Machu Picchu.

Peru's vast mineral, agricultural, and water resources have long been exploited by foreign companies and a few wealthy Peruvian families. The size of the nation is equal to the combined areas of Great Britain, France, and Spain. Nearly all of Peru's borders run across the high Andes mountains. The immense difficulties of travel posed by the Andes have impeded national unity. Iquitos, a river port on the Amazon River, lies only six hundred miles from the capital, but before the advent of the airplane, travelers between the cities were forced to choose either a direct and dangerous mountaineering trek, or a seven-thousand-mile circuitous trip that traversed the Amazon, the Atlantic, the Caribbean, the Isthmus of Panama, and the Pacific. As complicated as it may sound, this second option was the more frequent route.

Suriname

The capital of this nation, which lies on the northeastern coast just to the east of Guyana, is *Paramaribo*. Originally called Dutch Guiana, this country obtained its independence in 1975. Formerly a tropical colony with a plantation economy, now small farms have replaced the plantations, and Suriname has become one of the world's largest producers of bauxite (aluminum ore). The official language is Dutch, but English is more widely spoken.

The most striking characteristic of Suriname society is the heterogeneous ethnic composition of its population. In interracial communication, a language called Sranan is used—a mixture of English, Dutch, French, Spanish, and Hebrew! Each Indian tribe also has its own language, and Hindi, Javanese, Chinese, Lebanese, and other languages are spoken by other groups living in their own communities.

Uruguay

Positioned just below Brazil on the east coast with its capital at *Montevideo*, Uruguay has a predominantly white population. Most of the people are descendants of nineteenth- and twentieth-century immigrants from Spain, Italy, and other European countries. The Indian population is almost extinct. The language is a form of Spanish that has been influenced by Italian.

Venezuela

Venezuela is located on the northern coast of the continent, just to the east of Colombia. Its capital is *Caracas*. Tradition has it that the country was given its name, which means Little Venice, by Amerigo Vespucci, who was reminded of the Italian city when he first saw the native Indian houses that were built on stilts over the water. Until 1917, Venezuela was a poor and backward nation, but then oil was discovered, and the country became a leading fuel producer. It has since significantly developed its other industries and agriculture.

The Falkland Islands

This self-governing British colony is located in the South Atlantic Ocean, three hundred miles east of the southern tip of Argentina. The capital is *Port Stanley*. An English explorer made the first recorded landing in the Falklands in 1592, although the first settlers were French circa 1764. Since then, many claims have been made by Spain, Holland, France, and Argentina for control of the islands. The population is English-speaking, primarily of British descent, and presently numbers a little over two thousand. The whole area outside Stanley is devoted to sheep-raising, producing several million pounds of wool annually. All of it is sold to Great Britain. Argentina invaded and occupied the Falklands, which they call the Islas Malvinas, in April 1982 in order to end the "colonial situation." The Argentine garrison, however, was defeated by British forces six weeks later.

French Guiana

Situated on the northeast coast of the continent to the east of Suriname and north of Brazil, this region is a department of France and is represented in the French government by a senator and a deputy. But despite the French connection, the territory remains underpopulated and underdeveloped. There used to be a notorious penal colony in this dependency named Devil's Island—which was popularized in films and books like *Papillon*—but the place was closed in 1945 when France's convict settlements were abolished. Roads are found only in the coastal region, and the interior is largely uninhabited. The colony is economically dependent on France. Its capital is *Cayenne*.

THE LAND DOWN UNDER: AUSTRALIA

Australia, because of its large size, is considered a continent, *not* an island. Surrounded by oceans and seas, it's often designated as an "island continent" and is the smallest of all seven continents. The *nation* of Australia occupies the entire continent; its capital city is *Canberra*.

In geography discussions, Australia is often lumped together with many surrounding islands, such as New Zealand, New Guinea, and the Solomon Islands, and the entire region is designated as *Oceania*.

First settled as penal colonies for British convicts, Australia is now a prosperous, independent country, just a little smaller than the United States in land area. It has six states: New South Wales, Victoria, Queensland, South Australia, Western Australia, and Tasmania. The plains of Queensland and New South Wales support the world's greatest wool industry, and some of the most arid areas conceal great mineral wealth. The majority of Australians live on the coastal rim of the continent, which is hilly, fertile, and well-watered.

The most striking characteristics of Australia's three-million-square-mile land mass are its isolation, flatness, and dryness. From the air, this nation-continent looks like one huge desert. The inland part of Australia is called the Outback, which is the Australian equivalent of *frontier*. This land of hope and adventure is still sparsely populated and perhaps always will be.

Because Australia is remote from any other continent, it has many distinctive forms of plant life—notably the giant eucalyptus. There is also an unusual array of animal life, including the kangaroo, the flying opossum, the wallaby, the wombat, the platypus, the spiny anteater, the koala bear, and various striking birds. In 1788, rabbits were introduced into Australia, and they became a menace to sheep raising. In 1907, the rabbit population became such a problem that a fence one thousand miles long was built from the north coast to the south coast to prevent the rabbits from invading Western Australia.

There is a strong similarity in speech, manners, and customs throughout all the Australian states, and everywhere the culture of white Australia is immediately recognizable as similar to Anglo-Saxon culture in Britain and North America. Only a small percentage of the total population is nonwhite. This segment includes more than one hundred fifty thousand Aborigines, the earliest known inhabitants of Australia, who are thought to have migrated from Southeast Asia twenty thousand years ago. They live on reservations and have been classified anthropologically as a distinctive stock, the Australoid. These nomadic hunters and food gatherers are noted for their invention of the boomerang. The ancestry of present-day Australians has been estimated as about half English, one-fifth Irish, one-tenth Scottish, and a few Welsh. The remainder are largely from continental Europe. The most recent estimates put the total population at nearly eighteen million people.

The two major cities of Australia are Sydney and Melbourne:

Sydney: The capital of New South Wales, Sydney is the oldest and largest city in Australia—and is the only city in the country with a truly international atmosphere. Because of its magnificent harbor, it is one of the most important ports in the South Pacific. It's also the site of Australia's main international air terminal.

Melbourne: The capital of the state of Victoria, with a population of more than one million, this city is second in size only to Sydney. There is a good-natured rivalry between the two cities, with Melbourne having more of a reputation for conservatism and financial stability.

New Zealand

This island nation is, like Australia, an independent member of the British Commonwealth. New Zealand and Australia, along with Melanesia, are sometimes referred to together as Australasia, and they are part of the grouping of continent and islands that are known as Oceania. New Zealand is in a remote location, more than one thousand miles southeast of Australia. It comprises two main islands, the North and South Islands, and a number of smaller ones.

New Zealand was first settled by Polynesians, who remained isolated until the arrival of the European explorers. They had no name for themselves but eventually adopted the name Maori (meaning "normal") to distinguish themselves from the Europeans!

Known for its sheep industry and spectacular scenery, the country has been dependent economically on the export of agricultural products, especially to Great Britain. It's also begun to develop a more extensive and varied industrial sector to compensate for the fact that Britain will now have to turn to the European Community for many of its imports.

Wellington is the capital city, a major port, and the chief commercial center of the nation. *Auckland* is the largest city and chief port.

THE SOUTH POLE CONTINENT: ANTARCTICA

The name *Antarctica* means "opposite of the Arctic." This vast, ice-covered land on the southern tip of the globe is the fifth largest continent in terms of land mass. It's almost entirely covered by a vast sheet of ice averaging about sixty-five-hundred-feet thick, with some areas as deep as thirteen thousand feet thick! The winter temperatures range from minus 128 degrees Fahrenheit inland to minus 76 degrees near sea level. Around the coast, the ice sheets continually "calve," or discharge icebergs into the sea.

Antarctica is important as a region of international cooperation in scientific research. As a result, it's no longer the most unknown area in the world. The continent has been mapped and visited by many teams of geologists, geophysicists, glaciologists, and biologists. Sophisticated radio echo-sounding instruments are used to make aerial surveys of the land beneath the ice. Some of the below-ice formations are almost as well mapped as the exposed land.

Icebreakers and aircraft now make access to this frozen continent relatively easy—though there still can be some danger because of the harsh environment.

No description of Antarctica would be complete without reference to the heroic explorations of Captain Robert Scott, the British naval officer and Antarctic explorer. In 1910, he set forth for Antarctica in search of the South Pole. Scott reached a base on the Ross Sea in 1911, and in November of that year, he started southward on foot toward the pole. The explorer and his four companions pulled heavy sledges by hand across the high polar plateau in subzero weather. When they finally reached the South Pole on January 18, 1912, they found that the Norwegian explorer, Roald Amundsen, had beaten them to the spot by about one month! On their way back, this disappointed but heroic party was beset by illness, lack of food, frostbite, and blizzards. All five members died, the last three being overwhelmed by a blizzard when they were only a few miles from their destination. Their bodies were later recovered, along with Scott's diaries, records, and valuable scientific materials. His journey constitutes one of the epic, tragic events of exploration.

What's the future of this remote, cold continent? Antarctica was ice-free during most of its lengthy geologic history. Scientists believe there is reason to assume it may thaw out again in the very distant future—though when and how that might occur must remain a matter of speculation for now.

With this information about the continents and their nations in mind, we have a broad background to examine more closely some of the most important centers of human population—what I call the super cities.

CHAPTER EIGHT

THE SUPER CITIES

It's well and good to become adept at dealing with geography in the broad strokes of continents and nations. This background is essential for anyone who wants to become a really astute student of politics and history. But many times, especially when you're traveling to different parts of the world, a broad scope of knowledge must be accompanied by a focus on a more limited geographical area—especially on major cities. Those who know their cities are well on their way to becoming true geographical savants.

I believe that the first step in fleshing out your knowledge of specific urban areas should be to identify the most important cities you're most likely to have to know or visit. To get you started, I've chosen my own twenty "super cities," which currently play a major role in world affairs. First, the cities are listed. Then, I've included a quiz, with characteristics of each city, to challenge you to test your knowledge of these places.

After you've looked over these cities, I'd encourage you to add at least ten more that are especially important to *you,* either because of your work; because of a personal interest, such as your desire to return to them as vacation spots; or because they occupy a special position for you as a result of news reports or some other factor. Then, after you've gone over my twenty cities and added your ten, we'll move on to some memory devices for remembering state capitals.

The Twenty Super Cities

1. Paris
2. London
3. New York
4. Hong Kong
5. Rio de Janeiro
6. New Delhi
7. Singapore
8. Brussels
9. Rome
10. Moscow
11. Tokyo
12. Beijing
13. Athens
14. Amsterdam
15. Johannesburg
16. Berlin
17. Cairo
18. Washington
19. Geneva
20. Jerusalem

To help you understand the significance of these cities, here is a quiz indicating important characteristics of each. See if you can name the cities after reading each description.

Can you name the city *and* the country?

1. This city, formerly the capital, was divided into East and West after World War II and separated by a wall, which was pulled down in 1989. Now it is the capital again.

_____

2. A world financial, commercial, industrial, and cultural center. World sea port. Home of Westminster Abbey, Buckingham Palace,

Big Ben, Houses of Parliament. Established in A.D. 43 as Roman town of Londinium.

---

3. International cultural, intellectual, and fashion center, and commercial and industrial focus of the country. Tourist attractions include: Eiffel Tower, Louvre Museum, Notre Dame Cathedral. The Champs Elysees is the famous boulevard. This city grew from a pre-Roman settlement.

---

4. Cultural and business center and headquarters of several international organizations including the European Economic Community and NATO (North Atlantic Treaty Organization).

---

5. According to legend, this city was founded in 753 B.C. by Romulus. It is the seat of the Roman Catholic Church at Vatican City. One of the world's great centers of history, art, architecture, and religion. Called the Eternal City. Famous tourist landmarks: the Colosseum, Pantheon, Forum, St. Peter's Basilica in the Vatican, the Appian Way.

---

6. This city became capital of the country in 1340. In the middle of the city is the Kremlin and Red Square. Its attractions to visitors include the Tomb of Lenin, Gorki Park, and the Bolshoi Ballet Theater.

---

7. Founded in the twelfth century as a fortress for a warlord, it is now a world industrial hub. Its capital is one of the largest and most modern of cities. The bustling Ginza, a shopping and entertainment center, is well known to Westerners.

---

8. A major port and historical center, situated on the Nile delta, and founded on the site of a seventh-century military camp.

---

9. This city, known in the West for many years as Peking or Peiping, is the cultural, financial, and political center of the country. Renamed in about 1975, parts of it, known as the Forbidden City and

Imperial City, used to be the home of emperors, until it became communist headquarters.

_____

10. Founded in 1300, this city became home of one of the world's chief stock exchanges and a center of the diamond-cutting industry. It is also known for the Rijksmuseum (a great art museum) and for its canals and tulips.

_____

11. Founded in 1790, this seat of government of its country is known for the Lincoln Memorial, the White House, and other historical monuments.

_____

12. Founded by the Portuguese in 1502, this second-largest city in the country has a beautiful natural harbor. Among its famous landmarks are Sugar Loaf Mountain and Corcovado peak on which stands a colossal statue of Christ. Also known for its beaches, especially the Copacabaña, and its carnivals.

_____

13. Known as the Holy City, it is sacred to Christians, Moslems, and Jews alike. Especially noted for its shrines, including the Church of the Holy Sepulchre, The Dome of the Rock, the Western ("Wailing") Wall, and others. There is evidence that this city was occupied as far back as the fourth millennium B.C.

_____

14. This city is one of the world's biggest ports. It was destroyed in 1365, refounded in 1819, and was a British colony from 1946 to 1959, when it became independent. It is on an island at the southern tip of the Malaysian Peninsula.

_____

15. The largest city in this country and a center of the diamond and gold mining industries. This city (and entire country) is known for its racist policy of apartheid, which denies nonwhites equality with whites. New laws are now being negotiated.

_____

16. Site of the European headquarters of the United Nations and many other international organizations such as the World Health Organization and the International Red Cross. It is also very well known for its manufacture of watches and jewelry.

_____

17. The cultural legacy of this ancient city is incalculable. It is known to the Western World as the "cradle of democracy," and its historic monuments attract many thousands of tourists and students annually. The Parthenon is a worldwide symbol of this city.

_____

18. A British Crown colony located on the south coast of China, with its lease running out! The territory will return to Chinese rule in 1997. It is a free port, an international shipping center, and an important air hub.

_____

19. The largest city in the United States, one of the leading ports of the world, and a center for international air travel. It has vast cultural and educational resources, famous shops and restaurants, most of the nation's legitimate theaters, huge parks, and botanical gardens. It is known worldwide for many symbols of American life: Wall Street, Fifth Avenue, Broadway, Greenwich Village. It is a major tourist attraction.

_____

20. This city was built from 1912 to 1929 to replace Calcutta as the capital. It's the administrative center of the country. Mahatma Gandhi was assassinated there on a prayer ground in 1948.

_____

ANSWERS

The Super Cities

1. Berlin, Germany
2. London, England
3. Paris, France
4. Brussels, Belgium
5. Rome, Italy
6. Moscow, Russia
7. Tokyo, Japan
8. Cairo, Egypt
9. Beijing, China
10. Amsterdam, the Netherlands
11. Washington, DC, USA
12. Rio de Janeiro, Brazil
13. Jerusalem, Israel
14. Singapore, Republic of Singapore
15. Johannesburg, South Africa
16. Geneva, Switzerland
17. Athens, Greece
18. Hong Kong, China
19. New York, USA
20. New Delhi, India

Now that you know something about *my* twenty top cities, add ten more of your own. Jot them down on a separate sheet of paper, and add a few key facts about each one. You may want to gather these facts from an atlas, an encyclopedia, a tour guide, or just from your own recollections. When you've finished, insert your page at this point in the text so that you'll have it available for later reference.

Many times, I find that people who do this exercise return to their personal sheet in subsequent months and years and add to the list—an excellent idea, especially if you have some important personal or business need to jog your memory about your special cities.

Now, for the final section in this chapter, I encourage you to try what may seem to be something of an elementary-school assignment: memorizing the state capitals.

WHY MEMORIZE THE STATE CAPITALS?

At first, memorizing state capitals may seem a questionable or even useless exercise, but there are at least three reasons for becoming familiar with the fifty United States capitals:

- First, many of these cities are extremely important business, commercial, and political centers. The more of them you know, the stronger your grasp will be of the overall geography of the United States.
- Second, knowing the state capitals will give you a head start in conversations with friends and colleagues from other states. Many people from Illinois, for instance, may not expect out-of-staters to know that Springfield, rather than Chicago, is the capital. It will be a pleasant surprise and demonstrate more than a passing interest in their home area, if you indicate you know this fact.
- Third, learning the state capitals is sometimes an easy way to *appear* intelligent and knowledgeable.

  Surprisingly few people, for instance, know that Montpelier is the capital of Vermont; or that Pierre is the capital of South Dakota; or that Olympia is the capital of Washington. If questions about such capital cities arise in the course of a casual conversation, and you can rattle off the answers, you'll make an impression that most people won't forget.

Now, here are the fifty states and their capitals, with some tips for remembering them.

Alabama   The name of this city and one other state capital both begin with *Mont.* (*Montgomery.*)

Alaska   Sounds like the question, "Do you know?" (*Juneau.*)

Arizona   The name of a fantastic bird of mythology, which according to fable lived for five hundred years. After this enormous lifespan, the bird burned itself to ashes and emerged from the embers with renewed youth and beauty. (*Phoenix.*)

Arkansas   This capital became a symbol of the South's resistance to school integration in the 1950s, but the presidential campaign of Governor Bill Clinton helped restore its image in 1992. (*Little Rock.*)

California   Name of this city is from a Spanish word, derived from Latin, and even earlier from the Greek word for mystery. Also, think of a religious rite. (*Sacramento.*)

Colorado   Known as the "mile-high city" of the Rockies. (*Denver.*)

Connecticut   Center of the insurance business—an industry that can help "heal a broken *heart*" when it pays off a survivor. (*Hartford.*)

Delaware   There are no "white cliffs" near this city. (*Dover.*)

Florida   Do you see what I see? A Seminole Indian word. (*Tallahassee.*)

Georgia   Which city did Hollywood "burn" in a spectacular sequence in *Gone With the Wind*? (*Atlanta.*)

Hawaii   What's popular for leis, Waikiki, and nearby Pearl Harbor? (*Honolulu.*)

Idaho   It rhymes with "noisy." (*Boise.*)

Illinois   A capital that contains a season of the year. (*Springfield.*)

Indiana   The name of the state, plus the Greek for "city." It's known for its big auto races. (*Indianapolis.*)

Iowa   The name rhymes with *coin*. (*Des Moines.*)

Kansas.   This one is part of the name of a famous train that was glorified in an old song: "Acheson, ____ and the Santa Fe." (*Topeka.*)

Kentucky   Think: a straightforward fort. (*Frankfort.*)

Louisiana   French for "red stick." (*Baton Rouge.*)

Maine   This name is close to that of a Roman emperor. (*Augusta.*)

Maryland   The king of Siam—in *The King and I*—was in love with the woman mentioned in the first part of this city, with the Greek for *city* tacked on at the end. (*Annapolis.*)

Massachusetts   Sometimes known as "Beantown," this capital is the home of the Red Sox. (*Boston.*)

Michigan   Name suggests a knight's weapon. (*Lansing.*)

Minnesota   This is the smaller of the "twin cities." (*St. Paul.*)

Mississippi   Remember the name of the seventh president of United States? (*Jackson.*)

Missouri   Combine City with the name of the third president of the United States (*Jefferson City.*)

Montana   Think of Troy and the face that launched a thousand ships. (*Helena.*)

Nebraska   This capital is named for the sixteenth American president. It's also the brand name of a car. (*Lincoln.*)

Nevada   Named after a legendary Western figure whose first name was Kit. (*Carson City.*)

New York   It's *not* New York City! Instead, think in terms of Alba Milk and add NY for New York. (*Albany.*)

New Mexico   This capital has sort of a queer (clue) spelling and sound. (*Albuquerque.*)

New Hampshire   The name means mutual understanding. (*Concord.*)

New Jersey   The capital of New Jersey was the spot where Washington's forces first opened fire on the British in 1776. Its name may remind you that a "train weighs a ton." (*Trenton.*)

North Carolina Recalls the name of an English adventurer who is reputed to have placed his cloak over a muddy puddle for Queen Elizabeth I. (*Raleigh.*)

North Dakota This name is the same as that of the well-known nineteenth-century German statesman. (*Bismarck.*)

Ohio The last name of the explorer who opened the Western hemisphere to Europeans. (*Columbus*)

Oklahoma This okay state's name is also part of its capital's name. (*Oklahoma City*)

Oregon There are twelve cities in the United States with the same name—one of which is known for the Massachusetts witchcraft trials of the seventeenth century. (*Salem.*)

Pennsylvania It's *not* Philadelphia! Instead, think of a hairy burg. (*Harrisburg.*)

Rhode Island We often thank ———— when something good happens to us. (*Providence.*)

South Carolina Name of an Ivy League university and also a land that we hail in song. (*Columbia*).

South Dakota This capital is a common French first name, but locals pronounce it "peer." (*Pierre.*)

Tennessee You could grind your teeth trying to remember this town. (*Nashville.*)

Texas This is a brand of English automobile—and also the last name of a famous figure in Texas history (but don't mix it up with Houston!). (*Austin.*)

Utah The Dead Sea contains the same substance that can be found in the name of this capital. (*Salt Lake City.*)

Vermont What name takes its first syllable from French for *mountain* and is the other capital (besides Montgomery, Alabama) that begins with *Mont*? (*Montpelier.*)

Virginia The capital of the Confederacy during the Civil War. Its name will remind you of wealth. (*Richmond.*)

Washington Reminiscent of the mountain in Greece reputed to be the home of the gods. (*Olympia.*)

West Virginia Remember the name of a famous, fast dance of the 1920s and 1930s. (*Charleston.*)

Wisconsin Name of a famous avenue; the last name of President James ———; and of course, of his wife Dolley ———. (*Madison.*)

Wyoming   A name that suggests the Western frontier, cowboys, and an Indian tribe—but the people who live there aren't necessarily "shy." (*Cheyenne.*)

Now it's time to brush up on the new geographical lingo that you've been learning throughout this book. Check the following summary of what I call the language of location. After that, test what you've learned so far with the series of quizzes. And finally, resolve to use the concepts and techniques you've been assimilating as tools to *expand* your knowledge of the geographical dimensions of your daily life.

CHAPTER NINE

THE LANGUAGE OF LOCATION

You've learned a lot of new terminology so far in these pages. The following glossary is intended to refresh your memory on some of the most important terms. This special vocabulary comprises part of a "language of location" that should help you understand more fully the world in which we live.

GLOSSARY OF COMMON GEOGRAPHICAL TERMS

From *Helping Your Child Learn Geography*, published by the U.S. Department of Education, Office of Educational Research and Improvement, in cooperation with the Department of the Interior, U.S. Geological Survey.

Glossary

altitude
Distance above sea level.

atlas
A bound collection of maps.

archipelago
A group of islands or a sea studded with islands.

bay
A wide area of water extending into land from a sea or lake.

boundaries
Lines indicating the limits of countries, states, or other political jurisdictions.

canal
A man-made watercourse designed to carry goods or water.

canyon
A large but narrow gorge with steep sides.

cape (or point)
A piece of land extending into
water.

cartographer
A person who draws or makes
maps or charts.

continent
One of the large, continuous areas
of the earth into which the land
surface is divided.

degree
A unit of angular measure. A circle is divided into 360 degrees,
represented by the symbol °.
Degrees, when applied to the
roughly spherical shape of
the earth for geographic and
cartographic purposes, are each
divided into sixty minutes, represented by the symbol '.

delta
The fan-shaped area at the mouth,
or lower end, of a river, formed
by eroded material that has
been carried downstream and
dropped in quantities larger
than can be carried off by tides
or currents.

desert
A land area so dry that little or no
plant life can survive.

elevation
The altitude of an object, such as a
celestial body, above the horizon; or the raising of a portion
of the earth's crust relative to
its surroundings, as in a mountain range.

equator
An imaginary circle around the
earth halfway between the
North Pole and the South Pole;

the largest circumference of the
earth.

glacier
A large body of ice that moves
slowly down a mountainside
from highlands toward sea
level.

gulf
A large arm of an ocean or sea extending into a land mass.

hemisphere
Half of the earth, usually con-
ceived as resulting from the division of the globe into two
equal parts, north and south or
east and west.

ice shelf
A thick mass of ice extending from
a polar shore. The seaward
edge is afloat and sometimes
extends hundreds of miles out
to sea.

international date line
An imaginary line of longitude
generally 180 degrees east or
west of the prime meridian. The
date becomes one day earlier to
the east of the line.

island
An area of land, smaller than a
continent, completely surrounded by water.

isthmus
A narrow strip of land located between two bodies of water, connecting two larger land areas.

lagoon
A shallow area of water separated
from the ocean by a sandbank
or by a strip of low land.

lake
A body of fresh or salt water entirely surrounded by land.

latitude
The angular distance north or south of the equator, measured in degrees.

legend
A listing which contains symbols and other information about a map.

longitude
The angular distance east or west of the prime meridian, measured in degrees.

mountain
A high point of land rising steeply above its surroundings.

oasis
A spot in a desert made fertile by water.

ocean
The salt water surrounding the great land masses, and divided by the land masses into several distinct portions, each of which is called an ocean.

peak
The highest point of a mountain.

peninsula
A piece of land extending into the sea almost surrounded by water.

plain
A large area of land, either level or gently rolling, usually at low elevation.

plateau (or tableland)
An elevated area of mostly level land, sometimes containing deep canyons.

physical feature
A land shape formed by nature.

population
The number of people inhabiting a place

prime meridian
An imaginary line running from north to south through Greenwich, England, used as the reference point for longitude.

range (or mountain range)
A group or chain of high elevations.

reef
A chain of rocks, often coral, lying near the water surface.

reservoir
A man-made lake where water is kept for future use.

river
A stream, larger than a creek, generally flowing to another stream, a lake, or to the ocean.

scale
The relationship of the length between two points as shown on a map and the distance between the same two points on the earth.

sea level
The ocean surface; the mean level between high and low tides.

strait
A narrow body of water connecting two larger bodies of water.

swamp
A tract of permanently saturated low land, usually overgrown with vegetation. (A marsh is temporarily or periodically saturated.)

topography
The physical features of a place; or the study and depiction of physical features, including terrain relief.

valley
A relatively long, narrow land area lying between two areas of higher elevation, often containing a stream.

volcano
A vent in the earth's crust caused by molten rock coming to the surface and being ejected, sometimes violently.

waterfall
A sudden drop of a stream from a high level to a much lower level.

Glossary, in part, courtesty of Hammond, Incorporated

CHAPTER TEN

A GEOGRAPHY BEE

Now that you've finished an initial run-through of the geographical highlights that everyone should know, it's time to have a little fun. The following quizzes—which constitute a kind of geography bee—will test your knowledge and help you review some of the facts and concepts you may have forgotten from the previous pages. An answer section follows the quizzes. You'll most likely find new information in these questions and answers, so there will be an opportunity for your knowledge of geography to continue to grow.

After you finish these quizzes, use the questions contained in them as models to expand your understanding as you read various newspapers and magazines or watch television news reports. In other words, if you encounter a reference to an unfamiliar country or city, frame your own questions: "Where is this place? How many people live there? What are some of the basic features of its economy, politics, and historical background?" Read the article or listen to the news report with these queries in mind. If you don't find the answers in the report, pull out your atlas or encyclopedia. It won't take long to learn what you want to know—and you'll be well on your way to becoming a true geography virtuoso!

WHAT DO YOU KNOW ABOUT THE CONTINENTS: EUROPE*

1. The European continent is actually a vast peninsula of

---

* Answers on page 141.

2. What country is 3 percent in Europe and 97 percent in Asia?

_____

3. What is the name of the chain of mountains lying in south central Europe?

_____

4. A region in northern Europe containing Norway, Sweden, and Denmark is known as

_____

5. The capital of Finland is

_____

6. What is the capital city of Hungary?

_____

7. Which country is *not* in Europe?
Tunisia_____ Spain_____ Greece_____ Romania_____

8. What countries constitute the Iberian peninsula?

_____

9. What countries constitute Great Britain?

_____

10. What is the capital of Northern Ireland?

_____

11. What is the capital of the Republic of Ireland?

_____

12. The mountain chain between France and Spain is the

_____

13. The longest European river is the

_____

14. The capital of Wales is

_____

15. The capital of Switzerland is

_____

16. What is the name of the country (smaller than Washington, D.C.) that is located between Austria and Switzerland?

_____

17. What country lies south of the Alps?

_____

18. On what river is London located?

_____

19. What is the name of the Sea south of Europe and north of Africa?

_____

20. What country is known for its fjords?

_____

WHAT DO YOU KNOW ABOUT THE CONTINENTS: ASIA*

1. The Ural Mountains separate Asia from what other continent?

_____

2. In relation to size, how does Asia compare with the six other continents?

_____

3. Which country is *not* in Asia?
   China____ Japan____ Bulgaria____ India____
4. Calcutta is in
   India____ Bangladesh____ Pakistan____ Thailand____
5. The British Crown Colony of Hong Kong is in the Asiatic country of
   Japan____ China____ Burma____ India____
6. What is the name of the great mountain range lying in northern India and extending from Pakistan to China?

_____

7. What is the capital of Japan?

_____

8. What is the world's largest country in land area?

_____

9. What is the capital of India?

_____

* Answers on page 141.

10. Is Vietnam north or south of China?

_____

The Middle East is included in the lands of Asia. Here are ten more questions:

1. What is the predominant religion of the Middle East?

_____

2. What is the capital of Saudi Arabia?

_____

3. What nationality are the people on the island of Cyprus?

_____

4. What middle Eastern country lies partly in Europe?

_____

5. What country is Iraq's eastern neighbor?

_____

6. The world's first civilization was born in the fertile valley between two rivers in Iraq. Name the rivers.

_____

7. Name the holiest city of Islam.

_____

8. What country is at the southern end of Iraq on the Persian Gulf?

_____

9. What is the capital of Israel?

_____

10. In what country is the city of Beirut?

_____

WHAT DO YOU KNOW ABOUT THE CONTINENTS: AFRICA*

1. The world's largest desert is in Africa. What is its name?

_____

* Answers on page 141.

2. Name the large island in the Indian Ocean off the southeastern coast of Africa.

   _____

3. The longest river (four thousand miles) in the world is the

   _____

4. Name the capital of Sudan.

   _____

5. Addis Ababa is the capital city of

   _____

6. At Olduvai Gorge in the African country of _____,
   Louis B. Leakey, a British anthropologist, found the remains of the
   earliest-known humans.
7. Africa's highest mountain is

   _____

8. Name two capital cities of South Africa.

   _____

9. Mogadishu is the capital city of what country?

   _____

10. What country is known as the Gift of the Nile?

    _____

11. The countries of Morocco, Algeria, Tunisia, Libya, and Egypt are
    all bordered to the north by what body of water?

    _____

12. The western coast of the African continent is bordered by the

    _____ Ocean.
13. The largest city (not the capital) of Morocco is

    _____

14. The Belgian Congo achieved independence in 1960. Its name is
    now

    _____

15. If you sailed due west from anywhere in the African country of
    Angola, in what country would you land?

    _____

16. Where is the Cape of Good Hope?

_____

17. Name the capital of Kenya.

_____

18. What canal connects the Mediterranean and Red Seas?

_____

19. Eventually, part of eastern Africa may split off from the rest of the continent. This is being caused by huge plate movements in the

_____

20. What divides Morocco in North Africa from southern Spain?

_____

WHAT DO YOU KNOW ABOUT THE CONTINENTS: NORTH AMERICA*

1. What countries are included in the continent of North America?

_____

2. What country governs Greenland?

_____

3. Which of the following Canadian provinces does *not* border the United States?
   British Columbia____ Saskatchewan____ Ontario____
   Newfoundland____
4. Name the five Great Lakes.

_____

5. The official language of the Canadian province of Quebec is

_____

6. Name the capital of Canada.

_____

7. How many time zones are observed in the United States? What are they?

_____

8. In what hemisphere is the USA?

_____

* Answers on page 142.

9. Name the highest mountain peak in North America.

_____

10. The Mississippi River empties into the

_____

11. Name the two major rivers that flow into the Mississippi River.

_____

12. Which one of the Great Lakes is located entirely within the USA?

_____

13. Name the capital city of Mexico.

_____

14. In what country is the Yucatan Peninsula?

_____

15. Name three of the seven countries of Central America.

_____

16. What waterway connects the Atlantic and Pacific oceans?

_____

17. Name the huge bay in northern Canada.

_____

18. What is the name of the strait that separates Alaska from Siberia?

_____

19. On what island in the Hawaiian chain is Honolulu situated?

_____

20. Name the mountain chain in western North America.

_____

WHAT DO YOU KNOW ABOUT THE CONTINENTS: SOUTH AMERICA*

1. South America is divided into twelve independent countries. Name six.

_____

2. What is the name of the mountain chain that runs the entire length of South America?

_____

* Answers on page 142.

3. The largest country in South America is _____,
and its capital is

_____.

4. Name the capital of Argentina.

_____

5. What language is spoken in Brazil?

_____

6. One country has an unusual shape: one hundred miles wide (average) and twenty-seven hundred miles long. Name the country.

_____

7. The ancient Incan civilization once ruled in the region that is now the country of

_____

8. The world's second-longest river, after the Nile, is the

_____

9. Name the capital of Chile.

_____

10. What is the name of a group of islands off the coast of Argentina, claimed by Argentina but owned by Britain?

_____

11. South America is connected to North America by the

_____

12. Where is Cape Horn?

_____

13. In what country is Rio de Janeiro?

_____

14. Where is Caracas?

_____

15. Name the capital of Peru.

_____

16. Where is Bogotá?

_____

17. Two countries of South America are totally landlocked. Name one.

_____

18. Are the Andes mountains near the east or west coast of South America?

    _____

19. French Guiana is politically a part of France. Where in South America is it located?

    _____

20. Lying between Peru and Bolivia is the highest freshwater lake in the world. Its name is

    _____

WHAT DO YOU KNOW ABOUT THE CONTINENTS: AUSTRALIA AND NEW ZEALAND*

1. There are six states in Australia. Name three.

    _____

2. Name the capital of Australia.

    _____

3. Name the capital of New Zealand.

    _____

4. The inland part of Australia is called the

    _____

5. The original natives of Australia are the

    _____

6. Name Australia's largest city.

    _____

7. The Polynesian natives of New Zealand call themselves

    _____

8. The ancestry of most Australians is

    _____

9. Name three large Australian cities, other than Sydney and Canberra.

    _____

* Answers on page 143.

10. Because Australia is isolated from life on other continents, animals unique to Australia have developed there, such as

_____

WHAT DO YOU KNOW ABOUT THE CONTINENTS: ANTARCTICA*

1. Is Antarctica located around the North Pole or the South Pole?

_____

2. Do people live in Antarctica permanently?

_____

3. Is there any vegetation there?

_____

4. What is the Antarctic Circle?

_____

5. Around the Antarctic coast, the ice sheets continually "calve." What is meant by this term?

_____

WE'LL CLUE YOU—WHAT CITIES DO THE FOLLOWING BRING TO MIND?*

1. Broadway, Wall Street
2. Baked beans, well-known tea party
3. White House, Lincoln Memorial
4. Ship building, naval base
5. Liberty Bell, Declaration of Independence
6. Historical earthquake, Golden Gate Bridge
7. Mardi Gras, Creole culture
8. "Windy City," Al Capone
9. Entertainment center, smog
10. Television soap opera, oil wells

* Answers on page 143.

. . . AND WHAT STATES?

1. Grand Canyon, Painted Desert
2. Pilgrims, Lizzie Borden
3. Racehorses, Stephen Foster
4. Bering Sea, Arctic Ocean
5. Mormons, Salt Lake City
6. Jamestown, the "Virgin Queen"
7. Oranges, Disneyland
8. Oranges, Disney World
9. Hulas, leis
10. Yellow roses, lone stars

QUESTIONS WITH NO CLUES!*

1. A drawing that stands for something on a map is called a

_____

2. What is an inland body of water?

_____

3. What do we call a region of high, fairly flat land?

_____

4. When contour lines on a map are close together, they show that the slope of land is

_____

5. What is another name for any line of latitude?

_____

6. What is another name for any line of longitude?

_____

7. A chain of rocks lying just above or below the surface of the water is known as

_____

8. Symbols on a map are explained in the

_____

* Answers on page 143.

9. Which one of the following is *not* a natural feature?
   a lake____ a strait____ a canal____ an isthmus____

10. What is the difference between the mouth of a river and the source of a river?

   _____

11. A measure to show the relationship between a map and a real area on the earth's surface is known as a

   _____

12. We measure in feet above—and below—sea level, and we call this measurement

   _____

13. Physical or relief maps show

   _____

14. What do we call lines on a map that connect points of the same elevation?

   _____

15. Does the United States lie east or west of the Prime Meridian?

   _____

16. Where do lines of longitude meet?

   _____

17. Lines of latitude and lines of longitude combined form a pattern called a

   _____

18. Name the seven continents.

   _____

19. What is at 90°S?

   _____

20. The earth revolves upon an imaginary straight line. What is the name of this line?

   _____

HOW GOOD IS YOUR GEOGRAPHIC VOCABULARY?*

1. A peninsula is
   A) land nearly surrounded by water
   B) a small island near the mainland
   C) a sheltered place where ships can anchor safely
   D) an inland body of water

2. A sound is
   A) land near the mouth of a river
   B) a narrow strip of land connecting two larger land masses
   C) a body of water separating an island from the mainland
   D) part of a body of water reaching into the land

3. Maps and globes are divided into sections by imaginary lines called
   A) blocks
   B) grids
   C) contour lines
   D) tangents

4. Height of land above sea level is called
   A) latitude
   B) gradation
   C) longitude
   D) elevation

5. A relief map is a map showing
   A) political boundaries of countries
   B) physical features of the land
   C) historical data
   D) climatic conditions

6. Lines connecting points of the same elevation are known as
   A) location lines
   B) straight lines
   C) contour lines
   D) altitude lines

* Answers on page 144.

7. A plateau is
   A) a region of low flat land
   B) fertile land shaped like a basin
   C) a region of high, fairly flat land
   D) land at the peak of a mountain

8. An isthmus is
   A) a narrow strip of land connecting two larger land masses
   B) a narrow strip of water connecting two larger bodies of water
   C) an inlet of water into the land
   D) a canal dug through a narrow piece of land

9. A delta is
   A) a desert
   B) land deposited at the source of a river
   C) land deposited at the mouth of a river
   D) a small lake

10. A gulf is
    A) a large oil deposit
    B) a harbor
    C) a large deep arm of an ocean or sea extending into land
    D) an Arabian lake

11. A line of latitude is also called
    A) a line of longitude
    B) a meridian
    C) a parallel
    D) a time line

12. 0 latitude also means
    A) the prime meridian
    B) the Arctic Circle
    C) The Equator
    D) doesn't exist

13. What is 66.5 degrees north of the Equator?
    A) Antarctic Circle
    B) North Pole

C) Arctic Circle
D) the moon

14. Meridians are
    A) lines of longitude
    B) lines of latitude
    C) parallel lines
    D) astrological lines

15. In the time sense, P.M. means
    A) past midnight
    B) prime meridian
    C) post meridian
    D) post modern

16. What is the line of latitude at 23.5 degrees south?
    A) Tropic of Capricorn
    B) international date line
    C) 49th parallel
    D) Tropic of Cancer

17. In the time sense, A.M. means
    A) at midnight
    B) after midday
    C) ante meridian
    D) all meridians

18. Another name for the prime meridian is
    A) 0 degrees latitude
    B) 180 degrees latitude
    C) 0 degrees longitude
    D) 90 degrees longitude

19. The word *hemisphere* means
    A) a globe
    B) half a globe
    C) one quarter of a globe
    D) another word for *atmosphere*

20. An instrument with a needle used to tell direction is
    A) a weather vane
    B) a magnet
    C) a compass
    D) a Geiger counter

HOW IS YOUR GEOGRAPHIC KNOWLEDGE OF THE UNITED STATES?*

1. The Rocky Mountains are in the
   A) eastern part of the USA
   B) southeastern part of the USA
   C) central plains
   D) western part of the USA

2. What ocean or sea washes the eastern shores of the United States?
   A) North Sea
   B) Pacific Ocean
   C) Atlantic Ocean
   D) Caribbean Sea

3. Where is the Gulf of Mexico?
   A) northeast coast of USA
   B) southeast coast of USA
   C) west of Mexico
   D) near Hudson Bay

4. In relation to New York State, Alaska is
   A) southeast
   B) northeast
   C) northwest
   D) southwest

5. An example of a peninsula is
   A) Long Island
   B) Florida
   C) California
   D) Hawaii

* Answers on page 144.

6. The two states not connected to the continental United States are
   A) California and Rhode Island
   B) Alaska and Rhode Island
   C) Alaska and Hawaii
   D) Hawaii and Mexico

7. Where are the Appalachian Mountains?
   A) In the Rocky Mountain chain
   B) On the border with Canada
   C) In eastern United States
   D) In Texas

8. Which one of the following is *not* in the United States?
   A) Ohio
   B) Iowa
   C) Ontario
   D) Idaho

9. Which of the Great Lakes is entirely within the United States?
   A) Lake Huron
   B) Lake Michigan
   C) Lake Superior
   D) Lake Erie

10. Cape Cod is in
    A) Vermont
    B) Connecticut
    C) Massachusetts
    D) Maine

11. The capital of Texas is
    A) San Antonio
    B) Dallas
    C) Austin
    D) Houston

12. The United States part of Niagara Falls is in what state?
    A) New York
    B) Michigan
    C) Pennsylvania
    D) Vermont

13. Where is Cape Hatteras?
    A) Florida
    B) Massachusetts
    C) North Carolina
    D) Virginia

14. Which state of the union is nearest to Siberia?
    A) New York
    B) Hawaii
    C) Alaska
    D) Florida

15. About what city is it said, "It's a nice place to visit, but I wouldn't want to live there"?
    A) Chicago
    B) Cincinnati
    C) New York (borough of Manhattan)
    D) Las Vegas

16. Where is the Mojave Desert?
    A) New Mexico
    B) California
    C) Nevada
    D) Arizona

17. Mount Vernon, Virginia, was the home of
    A) George Washington
    B) Thomas Jefferson
    C) Alexander Hamilton
    D) Benjamin Franklin

18. South Dakota is the location of
    A) Mount McKinley
    B) Mount Rushmore
    C) Mount Saint Helens
    D) Mount Rainier

19. A city known for easy divorces is
    A) Reno
    B) Las Vegas

C) Los Angeles
D) New Orleans

20. The first permanent English settlement in North America was
    A) Boston, Massachusetts
    B) Jamestown, Virginia
    C) Richmond, Virginia
    D) Philadelphia, Pennsylvania

A GEOGRAPHY POTPOURRI*

1. Is Africa in the Eastern or Western Hemisphere?

   _____

2. Name four oceans of the world.

   _____

3. Where is the Baja Peninsula?

   _____

4. What is the meridian exactly opposite the prime meridian called?

   _____

5. Approximately how many miles is a degree of latitude on the surface of the earth?

   _____

6. How many degrees of latitude are there between 10 degrees north and 20 degrees south?

   _____

7. In what direction does a compass always point?

   _____

8. When contour lines are far apart, it indicates that the slope of the land is:
   steep____ gradual____ rough____
9. If a place is 10 degrees north of the equator and another place is *directly* south of the equator at 10 degrees, how many *miles* apart are they?

   _____

* Answers on page 144.

10. What is the latitude of the Arctic Circle?

_____

11. What is the capital of Scotland?

_____

12. The earth rotates on its axis once every _____ hours.
13. Halfway between the poles is an imaginary line of latitude called the

_____

14. Zero degrees longitude passes through a place called _____ in England.
15. How many degrees does the earth rotate in one hour?

_____

16. Latitudes between the Tropics of Cancer and Capricorn are known as the
    low latitudes_____ high latitudes_____ middle latitudes_____
17. What city is the capital of Oregon?

_____

18. Latitude 0 degrees divides the earth into two halves known as

_____

19. What is the measurement in degrees around the earth?

_____

20. What is the capital of Texas?

_____

ANSWERS

What Do You Know About the Continents: Europe (page 121)

1. Asia
2. Turkey
3. The Alps
4. Scandinavia
5. Helsinki
6. Budapest
7. Tunisia
8. Spain and Portugal
9. England, Scotland, Wales
10. Belfast
11. Dublin
12. Pyrenees
13. Volga
14. Cardiff
15. Bern
16. Liechtenstein
17. Italy
18. River Thames
19. Mediterranean
20. Norway

What Do You Know About the Continents: Asia (page 123)

1. Europe
2. It's the largest
3. Bulgaria
4. India
5. China
6. Himalayas
7. Tokyo
8. Russia
9. New Delhi
10. South

Middle East

1. Islam
2. Riyadh
3. Greek and Turkish
4. Turkey
5. Iran
6. Tigris and Euphrates
7. Mecca
8. Kuwait
9. Jerusalem
10. Lebanon

What Do You Know About the Continents: Africa (page 124)

1. Sahara
2. Madagascar
3. Nile River
4. Khartoum

141

5. Ethiopia (or Abyssinia)
6. Tanzania
7. Mount Kilimanjaro
8. Pretoria, Cape Town, Bloemfontein
9. Somalia
10. Egypt
11. Mediterranean Sea
12. Atlantic
13. Casablanca
14. Zaire
15. Brazil
16. Southern point in South Africa south of Cape Town
17. Nairobi
18. Suez Canal
19. Great Rift Valley
20. Straits of Gibraltar

What Do You Know About the Continents: North America (page 126)

1. Canada, USA, Mexico, Central America, and Greenland
2. Denmark
3. Newfoundland
4. Lake Huron, Lake Ontario, Lake Michigan, Lake Erie, and Lake Superior
5. French
6. Ottawa
7. Four: Eastern, Central, Mountain, and Pacific
8. Western (and Northern)
9. Mount McKinley, in Alaska
10. Gulf of Mexico
11. Ohio and Missouri
12. Lake Michigan
13. Mexico City
14. Mexico
15. Belize, Guatemela, Honduras, Nicaragua, El Salvador, Costa Rica, and Panama
16. Panama Canal
17. Hudson Bay
18. Bering Strait
19. Oahu
20. Rocky Mountains

What Do You Know About the Continents: South America (page 127)

1. Argentina, Bolivia, Brazil, Chile, Colombia, Ecuador, Guyana, Paraguay, Suriname, Peru, Uruguay, Venezuela
2. The Andes
3. Brazil, Brasilia
4. Buenos Aires
5. Portuguese
6. Chile
7. Peru
8. The Amazon
9. Santiago
10. Falkland Islands
11. Isthmus of Panama
12. Southern tip of South America
13. Brazil
14. Venezuela
15. Lima
16. Colombia
17. Bolivia and Paraguay
18. West
19. Northeast South America
20. Lake Titicaca

What Do You Know About the Continents: Australia and New Zealand (page 129)

1. New South Wales, Victoria, Queensland, South Australia, Western Australia, Tasmania (as well as the Northern Territory and the Australian Capital Territory)
2. Canberra
3. Wellington
4. The outback
5. Aborigines
6. Sydney
7. Maoris
8. British (English, Scottish, Irish, Welsh)
9. Melbourne, Hobart, Brisbane, Perth, Adelaide
10. Kangaroo, koala bear, platypus, Tasmanian devil

What Do You Know About the Continents: Antarctica (page 130)

1. South Pole
2. Only scientific researchers live there temporarily
3. No—Antarctica is almost completely covered with ice
4. An imaginary line of latitude at 66.5 degrees South
5. The ice sheets discharge icebergs into the sea

We'll Clue You—What Cities Do the Following Bring to Mind? (page 130)

1. New York
2. Boston
3. Washington, DC
4. Norfolk, Virginia
5. Philadelphia
6. San Francisco
7. New Orleans
8. Chicago
9. Los Angeles
10. Dallas

... and What States? (page 131)

1. Arizona
2. Massachusetts
3. Kentucky
4. Alaska
5. Utah
6. Virginia
7. California
8. Florida
9. Hawaii
10. Texas

Questions With No Clues (page 131)

1. Symbol
2. A lake
3. A plateau
4. Steep
5. A parallel
6. A meridian
7. A reef
8. Legend, or key
9. A canal
10. A mouth is the end, source is the beginning.
11. Map scale
12. The elevation
13. Physical features

14. Contour lines
15. West
16. At the Poles
17. Grid
18. Europe, Asia, Africa, North America, South America, Australia, Antarctica
19. The South Pole
20. The axis

How Good Is Your Geographic Vocabulary? (page 133)

1. A
2. C
3. B
4. D
5. B
6. C
7. C
8. A
9. C
10. C
11. C
12. C
13. C
14. A
15. C
16. A
17. C
18. C
19. B
20. C

How Is Your Geographic Knowledge of the United States? (page 136)

1. D
2. C
3. B
4. C
5. B
6. C
7. C
8. C
9. B
10. C
11. C
12. A
13. C
14. C
15. C
16. B
17. A
18. B
19. A
20. B

A Geography Potpourri! (page 139)

1. Eastern
2. Atlantic, Pacific, Indian, Arctic
3. Lower California, a part of Mexico
4. The International Date Line
5. Sixty-nine miles
6. 30 degrees
7. North
8. Gradual
9. 1,380 miles
10. 66.5 degrees north
11. Edinburgh
12. Twenty-four hours
13. Equator
14. Greenwich
15. 15 degrees
16. Low latitudes
17. Salem
18. Hemispheres
19. 360 degrees
20. Austin